A BRIEF
HISTORY
OF
SCIENCE

A BRIEF
HISTORY
OF
SCIENCE

CONSULTANT EDITOR

JOHN GRIBBIN

TED SMART

First published in 1998 by
WEIDENFELD & NICOLSON LTD
The Orion Publishing Group
5 Upper St. Martin's Lane
London WC2H 9EA

This edition produced for
THE BOOK PEOPLE LTD
Catteshall Manor, Catteshall Lane,
Godalming, Surrey GU7 1UU

A CIP catalogue record for this book is available
from the British Library

ISBN 0-297-82449-X

This book was conceived, designed and produced by
THE IVY PRESS LIMITED
2/3 St. Andrews Place
Lewes, East Sussex BN7 1UP

Art Director: PETER BRIDGEWATER
Editorial Director: SOPHIE COLLINS
Designer: JANE LANAWAY
Project Manager: JOHN WOODWARD
Project Editor: NICOLA YOUNG
Commissioning Editor: VIV CROOT
Page Make-up: CHRIS LANAWAY
Picture Research: VANESSA FLETCHER
Conceptualizer: STEPHEN OATES
Three-dimensional Models: MARK JAMIESON
Illustrations: MADELEINE HARDY, LORRAINE HARRISON, IVAN HISSEY,
SUZANNA HUBBARD, LESLEY ANN HUTCHINGS, ANDREW KAULMAN,
JOANNA KERR, JUDY STEVENS, SARAH YOUNG
Studio Photography: GUY RYECART, IAN PARSONS

Printed and bound in China

Contents

Science Now

INTRODUCTION

You could give Aristotle a tutorial – and you could thrill him to the core of his being. Aristotle was an encyclopedic polymath, an all-time intellect. Yet not only can you know more about the world than he did, you can also have a deeper understanding of how everything works. Such is the privilege of living after Newton, Darwin, Einstein, Planck, Watson, Crick and their colleagues.

I'm not saying that you're more intelligent than Aristotle, or wiser. For all I know, Aristotle was the cleverest person who ever lived. That's not the point. The point is only that science is cumulative, and we live later.

Aristotle had a lot to say about astronomy, biology and physics, but his views sound weirdly naive today. His ideas on other subjects sound fine, however. Aristotle could walk straight into a modern seminar on ethics, theology, political or moral philosophy, and contribute. But let him walk into a modern science class and he'd be a lost soul. Not because of the jargon, but because science advances, cumulatively.

Here's a sample of the things you could tell Aristotle, or any other Greek philosopher, and surprise and enthral them – not just with the facts but with how they hang together so elegantly.

The Earth is not the centre of the Universe. It orbits the Sun – which is just another star. There is no music of the spheres, yet the chemical elements from which all matter is made arrange themselves cyclically, in something like octaves. There are not four elements, but about a hundred – and earth, air, fire and water are not among them.

Living species are not isolated types with unchanging essences. Instead, over a timescale too long for humans to imagine, they split and diverge into new species, which then go on diverging further and further. For the first half of geological time our ancestors were bacteria. Most of the creatures on Earth still are bacteria, and each one of our trillions of cells is a colony of bacteria. Aristotle was a distant cousin to a squid, a closer cousin to a monkey, and a closer cousin still to an ape. (Strictly speaking, Aristotle was an ape, an African ape, a closer cousin to a chimpanzee than a chimp is to an orang-utan.)

The brain is not for cooling the blood. It's what you use to do your logic and your metaphysics. It's a three-dimensional maze of a million million nerve cells, each one drawn out like a wire to carry pulsed messages. If you laid all your brain cells end to end, they'd stretch around the world twenty-five times. There are about four million million connections in the tiny brain of a chaffinch, and proportionately more in ours.

If you're anything like me, you'll have mixed feelings about that recitation. On the one hand, pride in what Aristotle's species now knows and didn't then. On the other hand an uneasy feeling of, 'Isn't it all a bit complacent? What about our descendants, what will they be able to tell us?'

Yes, for sure, the process of accumulation doesn't stop with us. Two thousand years hence, ordinary people who have read a couple of books will be in a position to give a tutorial to today's Aristotles: to Francis Crick, say, or Stephen Hawking. So, will our view of the Universe turn out to be just as wrong?

Certainly there's much that we still don't know. But surely our belief that the Earth is round, and that it orbits the Sun, will never be superseded. That alone is enough to confound those who, endowed with a little philosophical learning, deny the possibility of objective truth: so-called Relativists who see no reason to prefer scientific views over aboriginal myths about the world. Our belief that we share ancestors with chimpanzees, and more distant ancestors with monkeys, will also never be superseded, although details of the timing may change.

On the other hand, many of our ideas are still best seen as theories or models, the predictions of which, so far, have survived the test. Physicists disagree over whether they are condemned forever to dig for deeper mysteries, or whether physics itself will come to an end in a final 'theory of everything', a nirvana of knowledge. Meanwhile, since there is so much that we don't yet know, we should loudly proclaim those things that we do understand, so as to focus attention on problems that we should be working on.

Far from being over-confident, many scientists believe that science advances only by disproof of its hypotheses. Konrad Lorenz said he hoped to disprove at least one of his own hypotheses every day before breakfast. That was absurd, especially coming from the grand old man of the science of ethology, but it is true that scientists, more than others, impress their peers by admitting their mistakes.

OPPOSITE **We can know more than Aristotle ever did and those that come after us will have an even greater knowledge about the world.**

A formative influence on my undergraduate self was the response of a respected elder statesmen of the Oxford Zoology Department when a visiting lecturer had just publicly disproved his favourite theory. The old man strode to the front of the lecture hall, shook the lecturer warmly by the hand and declared in ringing, emotional tones: 'My dear fellow, I wish to thank you. I have been wrong these fifteen years.' And we clapped our hands red. Can you imagine a politician being applauded for a similar admission?

Yet there is hostility towards science, hostility coupled with wilful ignorance. It has become almost a cliché to remark that nobody boasts ignorance of literature, but it is socially acceptable to boast ignorance of science and proudly claim incompetence in mathematics.

This is partly because people are suspicious of scientists. People certainly blame science for nuclear weapons and similar horrors. Yet who are the people who use these devices? It's been said before, but needs to be said again: if you want to do evil, science provides the most powerful weapons to do evil – but equally, if you want to do good, science puts into your hands the most powerful tools to do so. If we have the right aims, then science will provide us with the most effective methods of achieving them.

Then there's the view that science is dull, with rows of biros in its top pocket. According to one newspaper columnist and television critic: 'Science is constrained by experimental results and the tedious, plodding stepping stones of empiricism... What appears on television just is more exciting than what goes on in the back of it... There are stars and there are stars... Some are dull, repetitive squiggles on paper, and some are fabulous, witty, thought-provoking, incredibly popular...'

The 'dull, repetitive squiggles' is a reference to the discovery of pulsars in 1967, by Jocelyn Bell and Anthony Hewish. Jocelyn Bell Burnell had recounted on television the spine-tingling moment when, as a young woman on the threshold of a career, she first knew she was in the presence of something hitherto unheard-of in the Universe. Not something new under the sun, a whole new kind of Sun, which rotates so fast that, instead of taking twenty-four hours like our planet, it takes a quarter of a second. Is this dull?

Is it that science is just too difficult for some people, and therefore seems threatening? My own view is that the sciences can be intellectually demanding, but so can classics, so can history, so can philosophy. On the other hand, nobody should have trouble understanding things like the circulation of the

ABOVE **Modern science is based on theories developed by Isaac Newton in the seventeenth century. He discovered the law of gravity, laws of motion and how to use calculus. Many of his ideas were laid down in his book *The Mathematical Principles of Natural Philosophy*.**

blood and the heart's role in pumping it round. On one occasion John Carey, Merton Professor of English at the University of Oxford, quoted John Donne's lines to a class of thirty undergraduates in their final year reading English at Oxford:

Knows't thou how blood, which to the heart doth flow,
Doth from one ventricle to the other go?

Carey asked them if they knew how the blood does flow. None of the thirty could answer, although one tentatively guessed that it might flow 'by osmosis'. The truth – that the blood is pumped from ventricle to ventricle through at least fifty miles of intricately branched capillary vessels throughout the body –

Couldn't we treat science in the same way? Yes, we must have Bunsen burners and dissecting needles for those drawn to advanced scientific practice. But perhaps the rest of us could have classes in science appreciation, the wonder of science, scientific ways of thinking and the history of scientific ideas, rather than laboratory experience?

For science can be wonderful, inspiring and poetic. This may seem nonsense to some, who believe that science destroys the mystery on which poetry is thought to thrive. The poet Keats berated Newton for destroying the poetry of the rainbow:

Philosophy will clip an Angel's wings,
Conquer all mysteries by rule and line,
Empty the haunted air, and gnoméd mine –
Unweave a rainbow…

Blake, too, lamented:

For Bacon and Newton, sheath'd in dismal steel,
   their terrors hang
Like iron scourges over Albion; Reasonings like
   vast Serpents
Infold around my limbs…

I wish I could meet Keats or Blake to persuade them that mysteries don't lose their poetry because they are solved. Quite the contrary. The solution often turns out more beautiful than the puzzle, and anyway the solution uncovers deeper mystery. The rainbow's dissection into light of different wavelengths leads on to Maxwell's equations, and eventually to Einstein's special theory of relativity.

Einstein himself was openly ruled by an aesthetic scientific muse: 'The most beautiful thing we can experience is the mysterious. It is the source of all true art and science', he said. It's hard to find a modern particle physicist who doesn't own to some such aesthetic motivation. Typical is John Wheeler, one of the distinguished elder statesmen of American physics today:

…we will grasp the central idea of it all as so simple, so beautiful, so compelling that we will all say each to the other, 'Oh, how could it have been otherwise! How could we all have been so blind for so long!'

Wordsworth, unlike his fellow romantics, looked forward to a time when scientific discoveries would become 'proper objects of the poet's art'. At the painter Benjamin Haydon's dinner of 1817, he endeared himself to scientists by refusing to join in their toast, 'Confusion to mathematics and Newton'.

should fascinate any true literary scholar. So why the lack of interest? Could it be something to do with the way science is taught in schools?

Some time ago I received a letter which began, somewhat poignantly: 'I am a clarinet teacher whose only memory of science at school was a long period of studying the Bunsen burner.' Apparently the writer's interest in science was stifled by having to learn about the tools of the trade, rather than its achievements.

You can enjoy a clarinet concerto without being able to play the clarinet. You can even be a discerning and informed concert critic without being able to play a note. Clearly music would cease to exist if nobody learned to play it, but if everybody left school thinking they had to play an instrument before they could appreciate music, think how impoverished their lives would be.

Now, here's an apparent confusion. T. H. Huxley saw science as 'nothing but trained and organized common sense', while Professor Lewis Wolpert insists that it's deeply paradoxical and surprising, an affront to common sense rather than an extension of it. For example, every time you drink a glass of water, you are probably imbibing at least one atom that passed through the bladder of Aristotle. This is a tantalizingly surprising result, but in fact it is obtained by Huxley-style organized common sense, from Wolpert's observation that 'there are many more molecules in a glass of water than there are glasses of water in the sea'.

Science runs the gamut from the tantalizingly surprising to the deeply strange, and ideas don't come any stranger than quantum mechanics. More than one physicist has said something like 'If you think you understand quantum theory, you don't understand quantum theory.' There is mystery in the Universe, beguiling mystery, but it isn't capricious, whimsical or frivolous in its changeability. The Universe is an orderly place, governed by laws. If you put a brick on a table it stays there unless something lawfully moves it, even if, meanwhile, you forget it's there. Poltergeists and sprites don't intervene and hurl it about for reasons of mischief or caprice. There is mystery but not magic. There is strangeness beyond the wildest imagining, but no spells or witchery, no arbitrary miracles.

We are often urged to believe in such paranormal phenomena. Some people claim to be in touch with the dead, and others to prophesy the future. In one popular type of television programming, conjurers come on and do routine tricks. But instead of admitting that they are conjurers, these performers claim supernatural powers. Yet how can I be so confident that these 'supernaturalists' are ordinary conjurers?

It really comes down to parsimony, economy of explanation. It is possible that your car engine is powered by psychokinetic energy, but if it looks like a petrol engine, smells like a petrol engine and performs like a petrol engine, the sensible working hypothesis is that it is a petrol engine. There is nothing absolute about this judgement. A scientist would not rule out telepathy and possession by the spirits of the dead as a matter of principle. There is certainly nothing impossible about abduction by aliens in UFOs. One day it may happen. But on grounds of probability it should be kept as an explanation of last resort. It is unparsimonious, demanding more than routinely weak evidence before we should believe it. If you hear hooves clip-clopping down a street it could be a zebra or even a unicorn – but before you assume that it's anything other than

a horse, you should demand a minimal standard of evidence.

ABOVE **Astronomy, the realm of Yeats' 'starry ways', can fill us all with wonder, both as children and as adults**

It's been suggested that if the supernaturalists really had the powers they claim, they'd win the lottery every week. I prefer to point out that they could also win a Nobel Prize for discovering fundamental physical forces hitherto unknown to science. Either way, why are they wasting their talents doing party turns on television? By all means let's be open-minded, but not so open-minded that our brains drop out. Let us not go back to a dark age of superstition and unreason, a world in which every time you lose your keys you suspect poltergeists, demons or alien abduction.

The popularity of the paranormal, oddly enough, might turn out to be grounds for encouragement. I think that the appetite for mystery, the enthusiasm for that which we do not understand, is healthy and to be fostered. It is the same appetite which drives the best of true science, and it is an appetite which true science is best qualified to satisfy.

Astronomy, for example, is a science that deals almost exclusively with wonder and with the mysterious. To show how this sense of astronomical wonder can be presented to children, I'll borrow from a book called *Earthsearch* by John Cassidy.

Find a large open space and take a soccer ball to represent the Sun. Put the ball down and walk ten paces in a straight line. Stick a pin in the ground. The head of the pin stands for the planet Mercury. Take another nine paces beyond Mercury and put down a peppercorn to represent Venus. Seven paces on, drop another peppercorn for Earth. One inch away from Earth, another pinhead represents the Moon – the furthest place, remember, that man has explored. Fourteen more paces takes you to little Mars, then you must take 95 paces to reach giant Jupiter, a ping-pong ball. Add another 112 paces to reach Saturn, a marble. There is probably no space for the outer planets because the distances are much larger. But how far would you have to walk to reach the nearest star, Proxima Centauri? Pick up another soccer ball to represent it, and set off for a walk of 4200 miles (6760 km). As for the nearest other galaxy, Andromeda, don't even think about it!

Never mind the Universe, there is wonder to be found much closer to home. Your body contains a trillion copies of a large, textual document written in a highly accurate digital code, and each copy is as voluminous as a substantial book. I'm referring, of course, to the DNA in your cells. Textbooks describe DNA as a blueprint for a body. It is better seen as a recipe for making a body, because it is irreversible, but I want to present it as something different again, and even more intriguing. The DNA in you is a coded description of ancient worlds in which your ancestors lived.

The oldest human documents go back a few thousand years, originally written in pictures. Alphabets seem to have been invented about thirty-five centuries ago in the Middle East, and they've changed and spawned numerous varieties of alphabet since then. The DNA alphabet arose at least thirty-five million centuries ago, and since that time it hasn't changed one jot. The alphabet is not the only thing that has remained unchanged. The dictionary of sixty-four basic words and their meanings is the same, in bacteria and in us.

What changes are the long programs that natural selection has written using those sixty-four basic words. The messages that have come down to us are the ones that have survived millions, in some cases hundreds of millions, of generations. For every successful message that has reached the present, countless failures have fallen away like the chippings on a sculptor's floor. That's what Darwinian natural selection means. We are the descendants of a tiny élite of successful ancestors. Our DNA has proved itself successful, because it is here. Geological time has carved and sculpted our DNA to survive down to the present.

There are perhaps thirty million distinct species in the world today. This means that there are thirty million distinct ways of making a living: ways of working to pass DNA on to the future.

Some do it in the sea, some on land. Some up trees, some underground. Some are plants, using solar panels – we call them leaves – to trap energy. Some, the herbivores, eat the plants. Some eat the herbivores. Some are big carnivores that eat the small ones. Some live as parasites inside other bodies. Some live in hot springs. One species of small worm is said to live entirely inside German beer mats. All these different ways of making a living are just different tactics for passing on DNA. The differences are in the details.

The DNA of a camel was once in the sea, but it hasn't been there for a good 300 million years. It has spent most of recent geological history in deserts, programming bodies to withstand dust and conserve water. Like sandbluffs carved into fantastic shapes by the desert winds, camel DNA has been sculpted by survival in ancient deserts to yield modern camels.

At every stage of its geological apprenticeship, the DNA of a species has been honed and whittled, carved and re-jigged by selection in a succession of environments. If only we could read the language, the DNA of tuna and starfish would have 'sea' written into the text. The DNA of moles and earthworms would spell 'underground'. Of course all the DNA would spell many other things as well. Shark and cheetah DNA would spell 'hunt', as well as separate messages about sea and land.

We can't read these messages yet. Maybe we never shall, for their language is indirect, as befits a recipe rather than a reversible blueprint. But it's still true that our DNA is a coded description of the worlds in which our ancestors survived. We are walking archives of the African Pliocene, even of Devonian seas. You could spend a lifetime reading such messages and die unsated by the wonder of it.

There is an appetite for wonder, and true science is well qualified to feed it. It's often said that people need something more in their lives than just the material world. There is a gap that must be filled. People need to feel a sense of purpose. Well, not a bad purpose would be to find out what is already here, in the material world, before concluding that you need something more. How much more do you want? Just study what is, and you'll find that it already is far more uplifting than anything you could imagine needing.

You don't have to be a scientist – you don't have to play the Bunsen burner – in order to understand enough science to overtake your imagined need and fill that fancied gap. Science needs to be released from the lab into the culture.

This is an edited version of the Richard Dimbleby Memorial Lecture, given on BBC Television on 12 November 1996, with the title 'Science, Delusion and the Appetite for Wonder'.

	BC	BC	AD 1 – 799	AD 800 – 1499	1500 – 1599	1600 – 1699
MATHS	**530 BC** Pythagorus of Samos discovers the theorem about right-angled triangles named after him – the square on the hypotenuse is equal to the sum of the squares on the other two sides.	**300 BC** Euclid writes his *Elements of Geometry*, perhaps the most influential book in the history of mathematics and the basis of all geometry until the nineteenth century. *(EUCLID)*		**AD 825** Al-Khawarizmi writes the first major treatise on algebra, and introduces modern number notation based on "Arabic" numerals in which the value of the figures depends on their position.	*(ISAAC NEWTON)* **1614** John Napier invents the idea of logarithms and also "Napier's bones": a system of rods for making quick calculations.	**1687** Newton writes his *Principia Mathematica*.
PHYSICS	**250 BC** Archimedes of Syracuse founds the science of hydrostatics and discovers Archimedes' principle. This shows that a floating body experiences an upthrust equal to the weight of water it displaces.	*(ARCHIMEDES)*	**AD 50** Hero of Alexandria invents a fountain, a pump and even a steam turbine.	**AD 1000** Alhazen writes his book *The Treasury of Optics*, in which he describes lenses, mirrors and the camera obscura.	**1600** William Gilbert, in his book *De Magnete*, reveals that the Earth's magnetic field is like that of a giant bar magnet. *(NICOLAUS COPERNICUS)*	**1650** Otto von Guericke invents the air pump and investigates vacuums. He shows that when two large metal hemispheres are put together and the air within is pumped out, they cannot be separated even by teams of horses.
ASTRONOMY	**550 BC** Anaximander suggests that the Earth is a curved solid mass suspended in space.	**130 BC** Hipparchus of Rhodes makes the first systematic catalogue of stars, and develops a scale of magnitudes indicating their apparent brightness.	**AD 150** Ptolemy devises a complete model of the motion of the heavens based on orbits and epicycles. He also catalogues 1028 stars, and the sizes and the distances of the Sun and Moon.	**AD 980** Al-Sufi catalogues the position and brightness of over 1000 stars in his *Book of Fixed Stars*.	**1543** In his book *De Revolutionibus Orbium Coelestium*, Nicolaus Copernicus proposes that the Earth is not the fixed centre of the Universe, but circles around the Sun along with the other planets.	**1609** Johannes Kepler recognizes that planets follow elliptical, and not circular orbits, as expressed in his first two laws of planetary motion. His third law shows that the period of each planet's orbit is proportional to its distance from the Sun.	**1610** Galileo Galilei is the first to use a telescope to observe the heavens. He sees not only mountains on the Moon but, more significantly, moons orbiting Jupiter. This shows that an orbiting planet can carry moons with it, proving that the Earth can do the same.
CHEMISTRY	**450 BC** Empedocles suggests that all substances are formed from just four elements: fire, earth, air and water. These elements are joined or separated by two forces, attraction and repulsion – or more poetically, love and strife.	*(EARTH)*	*(AIR)*	*(FIRE)*	*(WATER)* **1556** Agricola writes his book *De Re Metallica*, which is not only a detailed record of sixteenth-century mining and metalworking, but also the definitive guide to the chemistry of the age.	**1661** Robert Boyle writes *The Skeptical Chymist*, in which he introduces the notion of elements and compounds and lays the foundations for modern chemistry. *(MOON)*
EARTH SCIENCES	**384–322 BC** Aristotle is one of the first to try to understand earthquakes scientifically. He believes that they are caused by air escaping from pockets where it is trapped underground. *(ARISTOTLE)*		**AD 78–139** Zhang Heng makes a seismoscope to record earthquakes.	**980–1037** Avicenna writes the book *Liber de Mineralibus* about various aspects of geology and earthquakes. This dominates thinking on mineralogy for 500 years.	*(AVICENNA)*	**1644** Torricelli realizes that the atmosphere possesses weight and exerts pressure. He proves it by creating the first mercury barometer – a glass tube full of mercury and sealed at one end.	**1646** Blaise Pascal demonstrates that air pressure decreases with altitude by taking a mercury barometer to the summit of Puy de Dôme, a 1200-metre mountain near Clermont Ferrand.
BIOLOGY	**330 BC** Theophrastus, considered to be the father of botany, accurately describes over 500 plant species and how they propagate.		**AD 83** Dioscorides writes *De Materia Medica*, which was to became the standard work on medicine for centuries.	*(DIOSCORIDES)*	**1543** Andreas Vesalius publishes his book *On the Structure of the Human Body* which completely supersedes Galen's ideas on anatomy – much to the annoyance of Galen's followers.	**1628** William Harvey's book *On the Motions of the Heart and Blood* describes for the first time the circulation of the blood and the role of valves in the heart, arteries and veins.	**1650** Marcello Malpighi discovers blood capillaries with the aid of the newly invented microscope.

1700 – 1799   **1800 – 1899**

1637
René Descartes creates coordinate geometry: the geometry of graphs, also known as Cartesian geometry, which changes geometrical problems into algebraic problems.

$$x^2 + y^2 = 1$$

RENÉ DESCARTES

1730
Leonhard Euler develops a huge number of theorems and puts trigonometry and differential calculus on a firm basis.

1755
Comte Joseph Lagrange begins his great book *Analytical Mechanics*, in which he uses the calculus of four-dimensional space to solve mechanical problems.

1791
Luigi Galvani discovers "animal electricity".

1768
Johann Lambert proves that the numbers π and e are irrational.

1799
Pierre Simon, Marquis de Laplace begins his work on celestial mechanics, the mathematics of the heavens.

THOMAS YOUNG

1801
Karl Gauss writes his book *Researches in Arithmetic*, which provides the basis of modern number theory and also gives the first proof that every natural number can be represented as the product of prime numbers in just one way.

1806
Jean-Robert Argand develops the Argand diagram, in which complex numbers are represented by points in the plane.

1665–1686
Isaac Newton introduces his theory of gravitation and three laws of motion, and also pioneers work on the nature of light.

1752
Benjamin Franklin uses a kite to show that lightning is electricity.

1803
Thomas Young provides convincing proof of the wave theory of light.

1820
Hans Oersted discovers that an electric current produces a magnetic field.

1665
Isaac Newton devises his theory of gravitation which explains Kepler's laws of planetary motion, and also explains the motion of the Moon, the Earth and the tides.

1705
Edmund Halley realizes that comets have predictable orbits and return at regular intervals. He successfully predicts the return, in 1759, of the comet that now bears his name.

URANUS

1781
William Herschel discovers the planet Uranus, the first "new" planet to be discovered since ancient times. He also discovered two moons of Uranus – Titania and Oberon – and two moons of Saturn, Mimas and Enceladus.

WAVE OF LIGHT

1801
Guiseppe Piazzi discovers the first asteroid, named Ceres.

1814
Josef von Fraunhofer analyzes the light from the Sun and shows that its spectrum contains dark lines. These "absorption lines" reveal a lot about the chemistry of the Sun's atmosphere.

1756
Joseph Black recognizes the importance of recording weight changes and the role of gases in chemical reactions. He also deduces the presence of carbon dioxide in the atmosphere.

1774
Joseph Priestley discovers a wide range of new gases, including nitrogen dioxide, ammonia, nitrogen, carbon monoxide, sulphur dioxide and oxygen.

1784
Henry Cavendish proves for the first time that water is not an element, when he explodes mixtures of hydrogen and air with an electric spark and creates water.

1789
Antoine Lavoisier writes his influential *Elementary Treatise on Chemistry* in which he defines a chemical element as the last point which analysis can reach. He also gives the first proper list of elements.

1800
Humphrey Davey discovers nitrous oxide – laughing gas – and suggests its use as an anaesthetic.

1808
John Dalton proposes his atomic theory, suggesting that every element is made of minute particles called atoms which can be neither divided or destroyed. According to his theory, every atom of each element is identical.

1669
Nicolaus Steno realizes that rocks form from sediment settling on the sea bed, and that the youngest layers will always be near the top – the principle of superposition.

1735
George Hadley suggests that warm air rising over the equator moves towards the poles in the upper air before cooling and sinking, identifying a circulation pattern now known as the Hadley cell.

1766
Jean-Étienne Guettard compiles the first geological map of France, and shows that there were once active volcanoes in the heart of France in the Massif Central.

1788
James Hutton is the first to realize that the Earth is many millions of years old, and that the landscape has been shaped gradually over very long periods.

1798
Henry Cavendish measures the density of the Earth from its gravitational effects and deduces that the Earth has a metal core.

1817
William Smith shows how the sequence of rocks within an outcrop can be dated from the fossils each layer contains.

JEAN LAMARCK

1830
Charles Lyell publishes his *Principles of Geology*, which becomes the geologists' standard text for more than a century. His key argument is that the world has been shaped gradually over geological time by the same slow forces operating today.

1658
Jan Swammerdam discovers red blood cells in the blood of a frog. He also helps pioneer the study of insects and their anatomy.

CARL LINNAEUS

1735
Carl Linnaeus makes the first great classification of plant species, grouping them into genera, orders and classes.

1757
Alexander Monro identifies the lymphatic system.

1767
Lazzaro Spallanzani, a pioneer in experimental physiology, demonstrates that life does not reproduce spontaneously by showing that a boiled broth that is hermetically sealed does not grow mould.

1801
Jean Lamarck proposes that animal species change through evolution in response to their environment.

1839
Theodor Schwann and Jakob Schleiden develop the cell theory of biology, which says that all plants and animals are made up of cells, each with a life of its own but each a part of the organism as a whole.

1800 – 1899 *continued*

MATHS

1822
Baron Joseph Fourier formulates partial differential equations to account for the heat flow in a solid body. To achieve this he develops the trigonometric series now called the Fourier series.

1826
Nikolai Lobachevsky shows for the first time in 2,000 years that there is another kind of geometry apart from Euclidian geometry – a fundamental discovery which later comes into its own with Einstein's general theory of relativity.

1829
Lambert Quételet develops statistical methods in his analysis of the Belgian population, and also shows how the theory of probabilities can be applied to the average man.

1833
Charles Babbage creates his "difference engine", and uses it to compile a table of logarithms. From this he develops the concept of his "analytical engine" which incorporates many of the principles of modern computers.

1854
Georg Riemann introduces the idea of Riemann surfaces and multi-dimensional space.

1880
Georg Cantor develops a highly original system of arithmetic for infinity.

GEORG CANTOR

PHYSICS

1830
Joseph Henry and Michael Faraday discover how a current can be magnetically induced - the basis of modern electricity generation.

1887
Albert Michelson and Edward Morley conduct an experiment showing that the speed of light is always the same – and that there is no such thing as the ether.

1888
Heinrich Hertz discovers radio waves.

1895
Wilhelm Röentgen discovers X-rays.

WILHELM RÖENTGEN

1896
Antoine Becquerel discovers radioactivity.

1897
J. J. Thomson discovers the electron.

1900
Max Planck devises the idea of quantum theory.

MAX PLANCK

1913
Ejnar Hertzsprung and Henry Russell discover the relationship between the colour of a star and its luminosity, devising the chart now known as the Hertzsprung-Russell diagram.

ASTRONOMY

1828
Caroline Herschel discovers eight new comets, and catalogues star clusters.

CAROLINE HERSCHEL

1838
Friedrich Bessel measures the distance to a star for the first time, using the parallax method to show that the star 61 Cygni is 10.3 light years away.

1843
Heinrich Schwabe discovers the 11-year cycle of sunspot activity.

1846
Johann Galle discovers the planet Neptune using the predictions of Urbain Le Verrier and John Couch Adams.

1862
Alvan Clark and his son discover that the Dog Star Sirius has a partner, Sirius B.

1896
Burkeland suggests that the aurora is created by charged solar radiation penetrating the Earth's atmosphere and becoming trapped in its magnetic field near the poles.

CHEMISTRY

1811
Amadeo Avogadro formulates Avogadro's law, which says that equal volumes of all gases contain the same number of "smallest particles" under the same conditions of temperature and pressure.

1818
Jöns Berzelius makes the first systematic table of relative atomic masses for the 28 elements then known. He also pioneers the idea of using initial letters as symbols for the elements.

1828
Friedrich Wöhler shows that organic chemicals are not unique to living things by synthesizing urea and pioneering organic chemistry.

1869
Dmitri Mendeléev creates the Periodic Table of chemical elements, in which the elements are placed (initially) in order of increasing relative atomic mass, and arranged into vertical groups possessing similar chemical characteristics.

1894
William Ramsay discovers the first of the noble gases, which came to be called argon. He later discovers other inert gases – krypton, xenon, neon and radon.

1898
Marie and Pierre Curie discover the radioactive elements polonium and radium. Marie Curie also shows that radioactivity is a natural property of the uranium atom.

EARTH SCIENCES

1863
John Tyndall explains for the first time why the sky is blue – because of the way particles in the air scatter blue light from the Sun. He also proposes the idea of the greenhouse effect.

1840
Louis Agassiz shows that at least once in the past much of the northern hemisphere was covered by a thick sheet of ice.

1857
Professor Buys Ballot shows that, in the northern hemisphere, winds blow anti-clockwise around centres of low pressure and clockwise around centres of high pressure. In the southern hemisphere the reverse is true.

1856
Luigi Palmieri realizes, while observing the eruption of Vesuvius in 1855, that an instrument that could detect very faint tremors in the ground might help predict earthquakes and volcanic eruptions. He builds the first modern seismograph.

1857
After studying the eruption of Vesuvius, Robert Mallet tries to work out how fast earthquake waves travel. He makes a map of earthquakes which shows, for the first time, that earthquakes occur in particular zones.

1874
Perrault makes the connection between rainfall and river flow, and finds that rivers are all supplied ultimately by rain and snow.

EARTH

1906
Richard Oldham shows that the Earth has a liquid core.

BIOLOGY

LOUIS PASTEUR

1856
Gregor Mendel discovers the basic statistical laws of heredity through his work on the edible pea.

1858
Charles Darwin and Alfred Wallace independently propose the theory of the evolution of species by natural selection.

1861
Louis Pasteur finally disproves the idea of spontaneous generation and suggests the germ theory of disease.

1863
Wilhelm Waldeyer-Hartz describes cancer as beginning with a single cell and spreading to other parts of the body by metastasis.

1867
Joseph Lister shows the value of antiseptic procedures during surgical operations.

1882
Ilya Mechnikov discovers the amoeba-like cells in the blood called phagocytes, which play a major part in fighting infection by digesting bacteria.

1902
Karl Landsteiner discovers that human blood can be divided into four groups: A, B, AB and O.

1900 – 1999

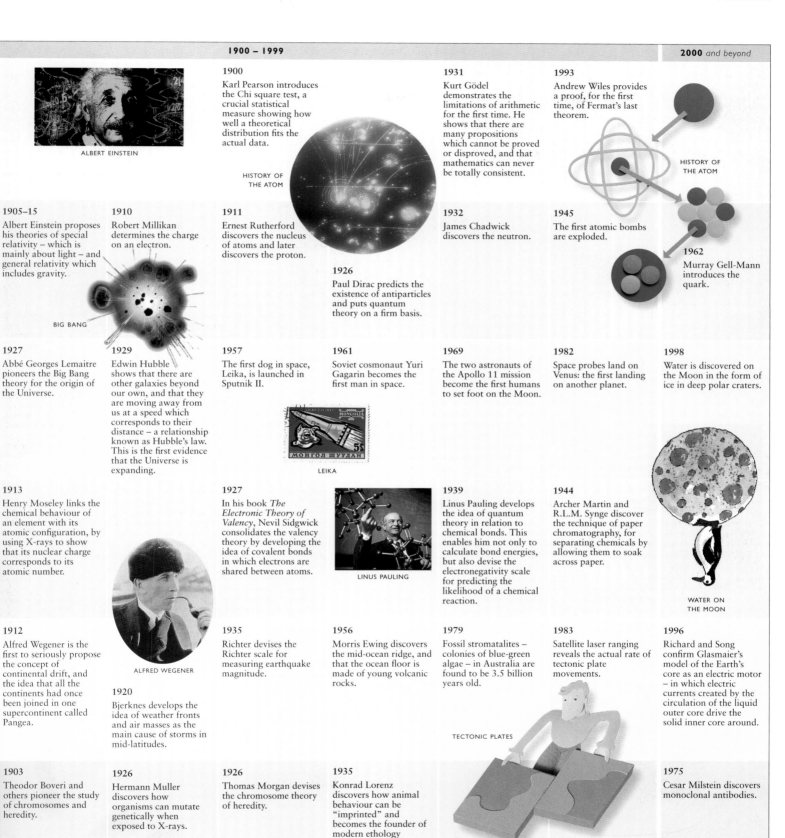

ALBERT EINSTEIN

1900
Karl Pearson introduces the Chi square test, a crucial statistical measure showing how well a theoretical distribution fits the actual data.

HISTORY OF THE ATOM

1931
Kurt Gödel demonstrates the limitations of arithmetic for the first time. He shows that there are many propositions which cannot be proved or disproved, and that mathematics can never be totally consistent.

1993
Andrew Wiles provides a proof, for the first time, of Fermat's last theorem.

HISTORY OF THE ATOM

1905–15
Albert Einstein proposes his theories of special relativity – which is mainly about light – and general relativity which includes gravity.

1910
Robert Millikan determines the charge on an electron.

1911
Ernest Rutherford discovers the nucleus of atoms and later discovers the proton.

1932
James Chadwick discovers the neutron.

1945
The first atomic bombs are exploded.

1962
Murray Gell-Mann introduces the quark.

BIG BANG

1926
Paul Dirac predicts the existence of antiparticles and puts quantum theory on a firm basis.

1927
Abbé Georges Lemaitre pioneers the Big Bang theory for the origin of the Universe.

1929
Edwin Hubble shows that there are other galaxies beyond our own, and that they are moving away from us at a speed which corresponds to their distance – a relationship known as Hubble's law. This is the first evidence that the Universe is expanding.

1957
The first dog in space, Leika, is launched in Sputnik II.

1961
Soviet cosmonaut Yuri Gagarin becomes the first man in space.

1969
The two astronauts of the Apollo 11 mission become the first humans to set foot on the Moon.

1982
Space probes land on Venus: the first landing on another planet.

1998
Water is discovered on the Moon in the form of ice in deep polar craters.

LEIKA

1913
Henry Moseley links the chemical behaviour of an element with its atomic configuration, by using X-rays to show that its nuclear charge corresponds to its atomic number.

1927
In his book *The Electronic Theory of Valency*, Nevil Sidgwick consolidates the valency theory by developing the idea of covalent bonds in which electrons are shared between atoms.

1939
Linus Pauling develops the idea of quantum theory in relation to chemical bonds. This enables him not only to calculate bond energies, but also devise the electronegativity scale for predicting the likelihood of a chemical reaction.

1944
Archer Martin and R.L.M. Synge discover the technique of paper chromatography, for separating chemicals by allowing them to soak across paper.

LINUS PAULING

ALFRED WEGENER

WATER ON THE MOON

1912
Alfred Wegener is the first to seriously propose the concept of continental drift, and the idea that all the continents had once been joined in one supercontinent called Pangea.

1935
Richter devises the Richter scale for measuring earthquake magnitude.

1956
Morris Ewing discovers the mid-ocean ridge, and that the ocean floor is made of young volcanic rocks.

1979
Fossil stromatalites – colonies of blue-green algae – in Australia are found to be 3.5 billion years old.

1983
Satellite laser ranging reveals the actual rate of tectonic plate movements.

1996
Richard and Song confirm Glasmaier's model of the Earth's core as an electric motor – in which electric currents created by the circulation of the liquid outer core drive the solid inner core around.

1920
Bjerknes develops the idea of weather fronts and air masses as the main cause of storms in mid-latitudes.

TECTONIC PLATES

1903
Theodor Boveri and others pioneer the study of chromosomes and heredity.

1926
Hermann Muller discovers how organisms can mutate genetically when exposed to X-rays.

1926
Thomas Morgan devises the chromosome theory of heredity.

1935
Konrad Lorenz discovers how animal behaviour can be "imprinted" and becomes the founder of modern ethology

1975
Cesar Milstein discovers monoclonal antibodies.

1952
Alfred Hershey and Martha Chase prove that DNA carries genetic information.

1953
Francis Crick and James Watson discover the double-helix structure of DNA.

FRUIT FLY

Chaos and Order

MATHEMATICS

We live on a digital planet. Our world is wired: you can pick up the phone and within seconds you may be talking to someone on the other side of the globe. The Earth wears a girdle of communication satellites which transmit thousands of television channels, millions of faxes, billions of phone calls every day. The Internet is only a modem away from any desktop computer. Soon all you will need is a bracelet on your wrist, through which you can talk to the World Information Network from the top of Mount Everest or the depths of the Amazon rainforest. The talkative ape is about to achieve its wildest dream: being able to talk to any other talkative ape, anywhere, at any time, about anything. And more than that: to be able to access, instantly, the entire combined archive of all previous talkative apes. You might call this 'extelligence'. It's like intelligence, but 'out there' in the culture. You want a book, a video, a film clip, a music track, a chemical formula, an engineering plan, a cave painting – push the right cultural buttons, and it's yours.

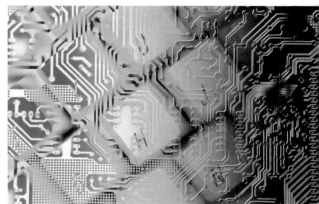

ABOVE **Without mathematics, the astonishing developments in micro-circuitry at the heart of computer and digital technology would not have been possible.**

BELOW **The information revolution will soon enable you to communicate with computers worldwide with just a wrist bracelet.**

This is the Information Age, or so we like to think. But it's more likely to be just the beginning. Information has become a commodity; in some respects information is more important than things, although this is often exaggerated. Raw information alone is worthless, as Internet users began to realize in the days before search engines. Even when information is all you need – and there are plenty of occasions when you need more, such as a CD-player to turn the information on a CD into music – you need to be able to find the information you want, and then make it do what you want it to. So what really counts is the algorithmic processing of information – pushing information through a system of rules to endow it with meaning.

The last thing that stockmarket traders need, for example, is more information. They already have far too much. What they need is a way to extract various useful features from the flood of raw information – patterns that indicate an impending change in the market trend, say. It's not enough to say 'put it on the computer'. It's already on the computer, but it's not there in a useful form. So the computer must be persuaded to process the information into something that can be used, and the way we do this right now is by algorithms – predefined, but often highly adaptable programs.

What lies at the heart of algorithmic processing of information? Something with a much longer history, going back to the days when the closest thing humans had to a computer was a pile of pebbles.

THE ROLE OF MATHEMATICS

The role of mathematics in our culture is distinctly curious. To most of us, mathematics is something that we were forced to do as children, didn't enjoy, weren't very good at, and still do our best to avoid. Many people break out into a cold sweat at the very thought of tackling a mathematical problem. And in everyday life we don't really see much evidence that our childhood tribulations were ever worth the effort.

This isn't helped by the fact that mathematics is used in three ways: working out mathematical formulae to find the answer to a specific problem (Calculus is used in this way); searching the body of existing mathematics for a solution to a new problem (a recent example of this is number theory, which is now used for coding messages on the internet, although it wasn't actually designed to do this) and finally an intellectual pursuit just for the fun of it! No wonder we were confused!

Evidence for the utility of art classes, on the other hand, is everywhere: the labels on cans, the posters on billboards, the advertising promotion for the latest blockbuster movie. The only mathematics we ever see is represented by the numbers on the till at the supermarket checkout, and the balance of the bank account – and we're mostly happy to take those on trust, even if we know we shouldn't.

The true role of mathematics lies far deeper. Mathematics brings order to a chaotic Universe: it provides us with the tools to dig below the surface of the Universe and find the secret rules that govern its behaviour. And paradoxically some of the most recent discoveries in mathematics have also revealed the secret chaos that underlies those rules. The Universe may well be stranger than we can imagine, but the extent to which our imagination can capture the Universe is growing explosively, and mathematics has played a major part in that explosion.

Even in our daily lives, mathematics is all around us. It is far more influential and important than, for instance, commercial art. Take away art, and the movie companies would find some other way to advertise their wares. Take away mathematics, and there would be no movies. Indeed, take away mathematics, and our civilization would collapse.

The reason that we don't usually witness mathematics in action is that – for entirely sensible reasons – it does its job behind the scenes. It is the part of the system that the user need not see; the hidden wiring that makes our civilization work, but would be too distracting (or frightening) if it were out in the open.

Say you go into a travel agent and book a flight to New York. The agent types a few instructions into a computer, and usually you don't even see what comes up on the screen. The only mathematics that you recognize is when you look at the bill. You are vaguely aware

ABOVE **As you feed in your request the computer uses mathematical logic to select search routes...**

BELOW **When you go into a travel agent to book your flight to Australia, or wherever you want to go on holiday, the computer that brings you instant booking information relies on mathematical processes to simplify the task...**

that more is going on, but you take it for granted and put it down to the marvels of the computer.

Let's pull back the curtain a bit, though. What is that computer doing? It's doing a lot more than just looking up a list of bookings and an airline timetable. It may be searching for a particular destination, to see which airlines fly there. But if it had to work its way through every airline, one by one, you'd be waiting all afternoon. Instead, it carries out the search in a clever way, to make it efficient. Searching is one of the simplest tasks in information processing. There are several favoured methods, and they're all based on mathematics.

But this is just the tip of an iceberg. How does the travel agent's computer communicate with the airline's computer? By telephone. What does that involve? A lot more than we usually think. Today our conversation over the telephone is split into thousands of very brief segments, and nearly all of those are promptly thrown away, leaving just a sample – say every hundredth segment. Such a

ABOVE **Within moments the booking information appears on the screen telling you availability...**

sample is enough to reconstitute the original sound, because human speech is very slow in comparison to the sampling rate: you're saying 'oooopppeeeennnn' and the sample is 'o p e n', so to speak, only more so.

After being sampled, each segment is 'digitized' or turned into a few numbers that define its important characteristics. The numbers from your segments are then interleaved with similar numbers from 99 other conversations. The result is snipped into relatively short 'packets' of information, and each makes its way independently through the telephone network. The packets are accompanied by 'routing' information that tells the phone system where to send them. When a packet gets to its destination, the numbers for your sample are stripped out and placed in a line, waiting their turn. As soon as the rest of your conversation up to that point has been dealt with, those numbers are turned back into sound in your friend's phone.

Without some very sophisticated mathematics, the system would never be able to keep track of where each bit of your conversation was, and how it should fit with the others. It wouldn't be able to shuffle your conversation in with the 99 others and get it back out again reliably. But because of the mathematics, the method works – and it squeezes 100 conversations into a phone line that used to take only one.

This is what it is like to live on a wired planet. This is how it works. The hidden layers of mathematics (and physics, of course, and chemistry, and biology, and economics, and art) can be peeled off almost forever. An onion is simplicity itself in comparison.

Some of the mathematics in those layers is very old indeed. The flights are run according to a schedule that is based on time. Sixty seconds to a minute, sixty minutes to an hour; why sixty? That numerical convention comes from the ancient Babylonians, 4,000 years ago. The Babylonians

BELOW CENTRE **The instant communication by phone right across the world which we now take for granted depends on a mathematical process called digital sampling. The phone splits your conversation into a few electronic number sequences that convey all the important characteristics of speech, and the sequences are then routed to their destination via a processing system that relies on some very clever maths.**

BELOW **Without mathematical shortcuts the computer would have to search through every airline and flight individually and the search would take all day.**

counted in sixties, whereas we now count in tens and hundreds (or twos, inside a computer). Nobody knows exactly why they did this – it may have been to do with astronomy (there are roughly 360 days in the year) or the fact that they were terribly pragmatic (60 is divisible by many numbers, including 1, 2, 3, 5, 6, 12, 15, 20 and 30, which is very useful).

The orbital dynamics of communications satellites relies on younger mathematics, a mere three centuries old. In 1687 Isaac Newton published the first edition of his *Mathematical Principles of Natural Philosophy*, in which he stated the law of gravitation (*see Chapter 2*). That's what you need to understand to determine satellite orbits and launch trajectories, along with Newton's laws of motion which tell you where the launch rocket will go.

Other mathematics is very new. The methods now used to schedule airline flights are less than ten years old. And if you take a camcorder with you on your holiday flight, it may well use 'fuzzy logic' to help keep the picture steady when your hand shakes. Fuzzy logic, a form of logic in which genuine half-truths exist (and are half true), was invented in the 1960s by Lotfi Zadeh (b. 1921 in Baku, Azerbaijan, now living in the United States). It hasn't yet become part of the mathematical mainstream, but Japanese engineers have taken to

RIGHT **People probably first began counting many thousands of years ago, but it took a long time before modern number notation developed.**

The earliest system was probably counting on fingers. But how could you record the number?

it like ducks to water, building it into hundreds of different consumer goods such as dishwashers and the ABS brake system now used in cars.

How did we reach this state? What is it that makes mathematics such a crucial part of human extelligence, such an important ingredient in our 'cultural capital'? Here are a few attempts that have been made, by various people, to describe or define mathematics:

- The science of pattern
- The science of significant form
- How to draw necessary conclusions
- The calculus of analogy
- The ultimate in technology transfer

They all say the same thing, really: that mathematics is our key to nature. It is a body of knowledge and technique that we human beings have evolved in order to describe, characterize, understand, manipulate, exploit, and control the world around us. It might turn out not to be the best way to achieve those aims – but right now it's the only thing we've got, and it works beautifully.

Why do so many of us dislike it? Perhaps because schools focus on technique, and good mathematical technique comes naturally to very few of us. But if you stop worrying about getting the sums right, relax, sit back, and enjoy the show, mathematics can be fascinating. It is humanity's way of bringing order out of chaos, of finding hidden patterns in places where all seems anarchy.

Let us begin by tracing just where mathematics came from, and how.

BELOW **Without elaborate mathematical routines the sophisticated electronic air traffic control systems that enable millions of people to fly around the world safely every day would be impossible.**

...then people learned to count
– and record – larger numbers
by dropping clay discs into a bag.

...then someone realized you
could simply make a scratch
on a tablet for each object.

...the **Babylonians** learned to
use different shaped marks on
clay to denote large numbers.

...now we use different symbols for
numbers up to nine, put in different
positions for larger numbers.

THE ORIGIN OF NUMBERS

Counting began long before there were symbols like 1, 2 and 3 for individual numbers. In fact you can count without using numbers at all. Counting on your fingers is a popular method. If you are a farmer you can work out that 'I have two hands and a thumb of cows' by folding down fingers and thumbs as your eye glances over the cows. You don't actually have to have the concept of the number 'eleven' to keep track of whether anybody is stealing your cows. You just have to notice that next time you seem to have only two hands of cows – so a thumb of cows has obviously gone missing.

You can also record the count as scratches on pieces of wood or bone, or you can make standard tokens to use as counters, such as clay discs with pictures of sheep on them for counting sheep, or pictures of cows for counting cows. As the animals parade past, you could drop the tokens into a bag, one for each animal. The use of symbols for numbers probably developed about 5,000 years ago when such counters were wrapped in a clay 'envelope'. It was a nuisance to break into the clay covering every time the accountant wanted to check the contents, and to make another one when he'd finished. So people put special marks on the outside of the envelope, summarizing what was inside. Then they realized that they didn't actually

EVOLUTION OF WESTERN NUMBER SYMBOLS

Hindu system

Arabic system

Spanish system

Italian system

0 1 2 3 4 5 6 7 8 9

need any counters inside at all: they could just make the same marks on clay tablets. It's amazing how long it can take to see the obvious.

The Babylonian number symbols, for instance, were made on pieces of wet clay with the corner of a stick. Not wanting to waste huge amounts of time making hundreds of such marks when the numbers got bigger, they invented a different shape for the number ten, and used multiples of that for 20, 30 and so on.

The number notation that we are familiar with today is quite different. Instead of repeating the same symbol many times to denote larger numbers, it uses a whole series of symbols: 0 1 2 3 4 5 6 7 8 9. On the other hand, instead of inventing new symbols for multiples of ten (such as % = 10, $ = 20, & = 30, @ = 40 and so on) it uses the same symbols 0 1 2 3 4 5 6 7 8 9 in different positions. We write the number 'five hundred and twenty-five' as 525, meaning $5 \times 100 + 2 \times 10 + 5 \times 1$.

ABOVE RIGHT **The modern number symbols evolved from the Hindu system of around** AD 800 **through the Arabic of** AD 900, **and the Spanish system of** AD 1000 **to the Italian system of** AD 1400 **which is similar to the system we now use today.**

This 'positional' way of writing numbers developed in India and Arabia between AD 200 and AD 800. The Hindu civilization in India goes back to about 2000 BC, and it developed mathematics from about 800 BC. Number symbols appeared there around 300 BC, a typical instance being the so-called Brahmi symbols. Brahmi notation differs from our present-day notation not just in the shapes of the symbols, but in the use of completely different symbols for 1 and 10, 2 and 20, and so on. However, it took a crucial step by employing nine different symbols for the digits 1 to 9.

By 600 AD, the Hindu system had developed into what was, apart from minor differences in the shapes of the basic symbols, today's decimal or 'base ten' number system. This means that it uses ten symbols (which happen to be 0 1 2 3 4 5 6 7 8 9) to represent the ten numbers from zero to

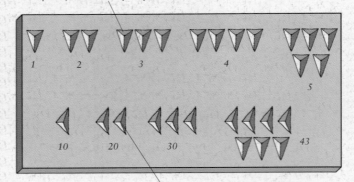

BABYLONIAN NUMBER SYSTEM

Multiples of the same shaped symbols were used for numbers 1 to 9

For numbers over 10 they used multiples of different symbols

ABOVE **In the Babylonian system all numbers are made up from just two symbols repeated – one for units and one for tens.**

nine. The next number, ten, is represented as the symbol 1 followed by the symbol 0, like this: 10. This is followed by 11, 12, 13 and so on up to 19; then we move on to 20, 21 and so on. Such two-digit combinations let us get as far as 99, but then we're stuck. So the next number, a hundred, is written as a string of three symbols, 100, and so on. The numbers that first require extra digits are 10, 100 and 1000: the powers of ten. It is important to have a symbol for zero, otherwise you can't reliably distinguish between numbers like 12, 102, 1002, 10200 and so on.

COUNTING IN BASE TWENTY

We've all got so used to the decimal system of notation that we tend to think that a number is the same thing as the series of symbols we use to write it down. But it isn't necessary to use number symbols remotely like ours. The Mayans, who lived in South America around AD 1000, worked to base twenty. In their system the symbols equivalent to our 525 would mean

$5 \times 400 + 2 \times 20 + 5 \times 1$

This comes out at 2,045 in our notation. The figure is so large because the first digit means 5×20^2 instead of 5×10^2, and $20^2 =$ 400. The actual symbols they used looked like this.

Early civilizations that used base ten probably did so because we have ten fingers. The Mayans may have counted on their toes as well, which is why they ended up inventing symbols using base twenty.

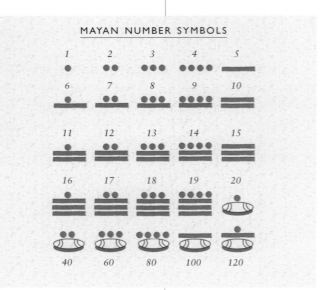

MAYAN NUMBER SYMBOLS

1	2	3	4	5
6	7	8	9	10
11	12	13	14	15
16	17	18	19	20
40	60	80	100	120

LEFT **Our number system uses a base ten, which means that it uses ten different number symbols. The Mayans of South America used a system of base twenty.**

THE IMPORTANCE OF ZERO

The history of zero goes hand-in-hand with that of 'place notation', the system that we now use to represent all whole numbers using just the ten symbols 0–9. This system relies on three ideas:

- A 'base' of ten, leading to the sequence units, tens, hundreds, thousands, and so on.
- Ten symbols to denote numbers from zero to nine.
- Using the position of a symbol to denote its numerical value in terms of the base. In the notation 1998, for instance, the first 9 means 'nine hundred' and the second 'ninety'.

Each of these ideas was around in very ancient times, but they first came together around AD 400 in the Hindu civilization in India, probably inspired by the abacus, a calculating device using beads that slide on wires or pebbles resting in grooves in the sand. It was realized that a row of beads could be represented by a symbol – not our current 1, 2, 3, 4, 5, 6, 7, 8, 9, but something similar. The symbol 9, say, could represent nine beads in any row – nine thousands, nine hundreds, nine tens, nine units. The symbol's shape wouldn't tell us which, but its position would. Place notation makes it relatively easy to do arithmetic on paper, slates, or clay tablets, without using an abacus at all.

ABOVE **Discovering the usefulness of 'zero' was one of the key advances in mathematics.**

When place notation was first developed, there was no symbol for 0. In the earliest recorded reference to Hindu numerals, written in the year 662, the Syrian bishop Severus Sebokt stated that 'computation is done by means of nine signs'. However, when using place notation, it is important to have a symbol for an empty row of beads: without it, you can't tell the difference between 14, 104, or 1040. So a tenth symbol, zero, was added to the list of digits. The first known Hindu reference to zero occurs in an inscription of 876, but the idea surely arose earlier.

When the symbol for zero was first introduced, zero wasn't considered to be a number in its own right: it was just an auxiliary symbol used in the representation of numbers. We don't know who first considered zero to be a number, but by 800 the Indian mathematician Mahavira was explaining that multiplying a number by 0 gives 0 and that subtracting 0 from a number yields the same number so 0 is a number like any other. In modern times, zero has become indispensable. Your bank's computer treats the money in your account as a number, and 0 has a very special significance! For an aeroplane, an altitude of zero indicates that it is on the ground.

GREEK GEOMETRY

Mathematically speaking, numbers come before more complex ideas such as algebra (the use of symbols to represent unknown numbers) and geometry (the mathematics of shapes). Historically, however, the first real flowering of mathematics occurred much earlier than the development of today's number system. It took place in Greece, between about 600 BC and 150 BC, and its focus was geometry. But in some ways its deepest contribution was logic, for the Greeks were the first to emphasize the notion of *proof*. Until then, mathematical truths were established by a mixture of argument and experiment. An example of such a truth being that the angles of a triangle add up to 180 degrees. From the time of the Greeks onwards, no mathematical statement could be considered established unless it had been given a logically rigorous proof.

The emphasis on proof, now felt to be the very essence of mathematics, evolved gradually and erratically. The earliest Greek school was that of Thales, who lived some time around 640–550 BC. Thales was reputedly an excellent businessman: one year he cornered all the olive presses and rented them out at a big profit. He is said to have calculated the height of the pyramids by comparing their shadows to those of a stick; he may even have predicted an eclipse of the Sun in 585 BC. And it is possible – although open to dispute – that Thales was the first Greek to provide proofs for some of his theorems.

Far more important, though, were the Pythagoreans, a semi-mystical cult which flourished between about 585 BC and 400 BC. The name of Pythagoras is known to every school student thanks to his famous theorem about right-angled triangles (the square on the hypothenuse is equal to the sum of the squares on the other two sides). The Babylonians used the theorem empirically (examples have been found on clay tablets dating from this time) although we have no idea what they used it for. It is also not known whether Pythagoras proved it, let alone whether his was the first proof.

TETRAHEDRON

THE CULT OF PYTHAGORAS

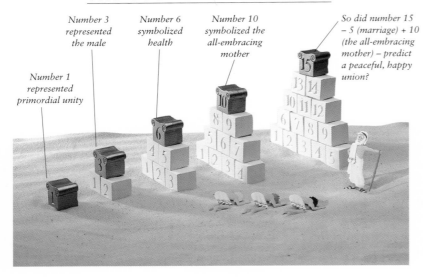

Number 1 represented primordial unity

Number 3 represented the male

Number 6 symbolized health

Number 10 symbolized the all-embracing mother

So did number 15 – 5 (marriage) + 10 (the all-embracing mother) – predict a peaceful, happy union?

The Pythagoreans saw the number 1 as the primordial unity from which all else is created, 2 as the symbol for the female, and 3 for the male. The number 4 was symbolic of harmony, because 4 = 2 × 2 and since 2 is even, so 4 is 'evenly even'; it also symbolized the four elements out of which everything in the Universe was thought to be made: earth, air, fire and water. The number 5, the sum of male 2 and female 3, symbolized marriage. The number 6 symbolized health, and was also 'perfect' because its divisors 1, 2, and 3 add up to 6 again.

The Pythagoreans got excited about 10, because as 1 + 2 + 3 + 4 it was the all-embracing mother, a symbol of completeness which combined the primordial unity, the female principle, the male principle, and the four elements.

The term 'square' for some numbers still survives today, and so does 'cube' for the analogous three-dimensional construction. For example, 3 cubed is 3 × 3 × 3 = 27.

PYTHAGORAS' TRIANGLE

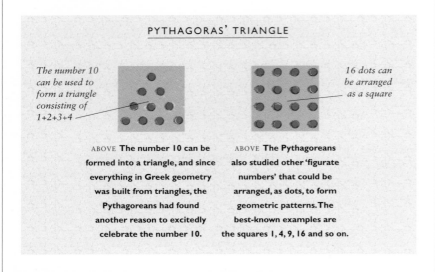

The number 10 can be used to form a triangle consisting of 1+2+3+4

16 dots can be arranged as a square

ABOVE **The number 10 can be formed into a triangle, and since everything in Greek geometry was built from triangles, the Pythagoreans had found another reason to excitedly celebrate the number 10.**

ABOVE **The Pythagoreans also studied other 'figurate numbers' that could be arranged, as dots, to form geometric patterns. The best-known examples are the squares 1, 4, 9, 16 and so on.**

Little is known about Pythagoras personally, except that he was born on the island of Samos; however, Pythagoras' cult changed the face of mathematics, because the Pythagoreans recognized that mathematics is about abstract ideas, not real things. They believed that the entire Universe was founded on numbers, and saw a deep meaning in the symmetries and sequences they formed. Such ideas were part philosophy, part mysticism, but they were founded on mathematical truths that inspired many of the Greek mathematicians who were to follow.

The best-known of the Greek mathematicians is Euclid, who probably learned his geometry in Plato's academy, lived in Alexandria from about 330 BC to 275 BC, and trained students. That's pretty much all we know about him as a person, but we know a lot more about his mathematics because an unusually large amount of it has survived the ravages of time, war and religious bigotry. No original manuscripts are known, but Euclid was widely copied.

His best-known work, the *Elements*, is a systematized collection of the geometric discoveries of his predecessors, no doubt augmented with many of Euclid's own. It comprises 13 separate books, and it makes explicit what later became known as the axiomatic method. This can be summarized as 'begin by stating your assumptions clearly, and deduce everything else from them with as much logical rigour as the prevailing mathematical culture demands'. Euclid

ICOSAHEDRON

TOP CENTRE **One of the great breakthroughs of classical mathematics was Archimedes' discovery of how to calculate the volume of a sphere.**

π or pi is the ratio of the circumference of a circle (approximately 3.14)

r is the radius of a sphere

the volume of a sphere is $^4/_3 \times \pi \times r^3$

CUBE

started from simple properties of lines and circles, and ended by proving that only five regularly shaped solids can possibly exist – the *tetrahedron*, *cube*, *octahedron*, *dodecahedron*, and *icosahedron*. The conceptually deepest part of the *Elements* occurs in Book V, which presents the ideas of another Greek, Eudoxus, about what are now called irrational numbers. These are numbers that cannot be written as a fraction $^p/q$, where p and q are whole numbers. In the spirit of Greek geometry Eudoxus actually worked with proportions between lines, not numbers as such. Books VII-IX of the *Elements* are also

noteworthy: they deal with the theory of numbers, proving in particular that there exist infinitely many prime numbers. (A number is prime if it has no divisors other than itself and one: the first few primes are 2, 3, 5, 7, 11, 13 and 17). Prime numbers are very important in mathematics – you could think of them as the lego blocks of natural numbers.

The other Greek geometer whose work has survived in substantial quantity is Apollonius who lived from about 262–190 BC. His greatest work is the *Conic Sections*. In it he derived fundamental properties of curves such as the ellipse, parabola and hyperbola, which are formed when a plane cuts a cone. The ellipse later proved crucial to the understanding of planetary orbits, but in Apollonius' day the value of conic sections was purely intellectual.

Archimedes (287–212 BC) was an astronomer's son born in Syracuse – an inventor as well as a mathematician. He found the areas and volumes of many shapes and showed that the volume of a sphere is $^4/_3\pi r^3$, where r is the radius and π is the ratio of the circumference of a circle to its diameter (roughly 3.14). His

OCTAHEDRON

proof related the volume of the sphere to an enclosing cylinder. The Greeks had no number for π, because their geometry rested on proportions, not numbers. Yet they knew it cropped up in the circumference and area of a circle, and in the volume and surface area of a sphere. Archimedes found an excellent approximation to π, proving that it lies between $^{223}/_{71}$ and $^{22}/_7$.

RATIONAL AND IRRATIONAL NUMBERS

Numbers like $^2/_3$, which can be written as a fraction, and whose numerator and denominator are whole numbers, are said to be rational. The Ancient Greeks discovered that numbers – such as the square root of two (which we now write as √2) – can exist that are not of this form.

In the 5th century BC a Greek called Hippasus of Metapontum proved, in geometric form, that no rational number, multiplied by itself, can give 2 exactly. In other words, √2 is irrational. This discovery had serious implications for Greek geometry. For example, take a square whose sides are of length 1. Pythagoras's Theorem implies that its diagonal has length √2. Since √2 is irrational, the side and diagonal of the square are *incommensurable*: they cannot both be whole number multiples of some common length. Many theorems that have easy proofs when lengths are commensurable become very difficult to prove if those length are incommensurable, and so the Greeks had to revise their methods.

The big breakthrough was made by Eudoxus (c. 408–355 BC), a Greek geometer, astronomer, physician, geographer and legistlator. It involved abandoning number as the basic concept and working directly with geometrical lines. In the long run, his method (which is highly technical) proved too clumsy to survive, and instead the concept of number was extended to permit irrationals. Even so, traces of the difficulties remain: e.g. if we write √2 as a decimal then the sequence of digits goes on forever and never settles into a repetitive cycle.

GEOMETRIC SHAPES **One of the first landmarks in mathematics was the Ancient Greek mathematician Euclid's discoveries about regular solids. Starting from the simple** properties of lines and circles, he went on to deduce that there are just five regular solids – the tetrahedron, the cube, the octahedron, the dodecahedron and the icosahedron.

DODECAHEDRON

THE MIDDLE AGES

In Europe the Middle Ages were notable for their almost total lack of mathematical innovation. This was probably due to the attitudes of the Roman church, which stifled new ideas. Therefore the serious action shifted to the Orient. By the year 500 the Indian mathematician Aryabhata had improved on Archimedes by showing that π is approximate to 3.1416. Brahmagupta, in about 625, gave the first method for finding integer (whole number) solutions of equations such as $5x + 7y = 29$. For example, here the only solution in positive integers is $x = 3$, $y = 2$. A Hindu mathematician, Bhaskara (1114–c.1185), in about 1150, allowed solutions to be negative, but was sceptical that negative numbers were meaningful.

During this time the decimal system of notation was developed, mainly by the Hindus. The earliest reference to this system is found in the writings of a Syrian bishop, Severus Sebohkt, in 662.

There was also plenty of mathematical activity in the Islamic world during the medieval period. The best-known Islamic mathematician was Muhammad al-Khwarizmi, whose name gave rise to the term 'algorithm'. Al-Khwarizmi discussed linear equations, which in modern notation take the form $ax + b = 0$, x being the unknown and a and b numerical coefficients. He also considered quadratic equations such as $ax^2 + bx + c = 0$. He did not employ any overt symbolism – indeed, geometry was used for what would now be algebra. He also made tables of trigonometric functions such as the sine and tangent.

Another first-rate Islamic mathematician was Omar Khayyám (about 1038–1123), although today he is best known for the Fitzgerald translation of his poem the *Rubaiyat*. Khayyám also discovered how to solve cubic equations such as $ax^3 + bx^2 + cx + d = 0$. The third great source of

new mathematics during the Middle Ages was China. Tsu Ch'ung Chi (430–501) discovered an approximation accurate to nine decimal places! Ch'in Chiu Shao devised a numerical method – now known in the west as Horner's method after William Horner who rediscovered it in 1819 – to approximate the roots of any algebraic equation.

EUROPEAN REVIVAL

Hindu-Arabic numerals were brought to Europe in 1202 by Leonardo of Pisa in his *Liber Abaci* – most of which was about trading on the foreign exchange markets. In 1220 Leonardo published his *Practica Geometriae*, which applied the Hindu-Arabic numerals to geometry and trigonometry (the mathematics of triangles – the most practical of applications which is used in surveying and navigating). He also invented a new kind of special number through considering a problem about rabbits.

Imagine that rabbits can breed in the second month of their life, and that each pair then produces a new pair of offspring each month. Starting with one pair of new born rabbits, how many pairs of rabbits will there be at the start of each month? What is the pattern of the number of pairs? Leonardo deduced that it is the sequence 1, 1, 2, 3, 5, 8, 13, 21, 34, 55, 89 etc., where each number is the sum of the two preceding ones. These are now called Fibonacci numbers, after a nickname given to Leonardo in the eighteenth century by french mathematician Guillaume Libri. They still hold a great fascination for mathematicians and are important in many areas of modern mathematics, especially dynamics and computer science. Incidentally, Libri was the writer of pop mathematics books during this period (a bit like this one!) and he probably gave Leonardo the nickname, Fibonacci (which means the son of Bonaccio or the Bonaccio Kid) to make him seem more hip to the readers.

GALILEO GALILEI
1564–1642
Galileo provided the first great insights into the mathematics of motion – recognizing the importance of acceleration. He also showed that the path of a projectile through the air is a parabola.

Interest in mathematics began to blossom in Renaissance Europe. In 1494 Luca Pacioli published his *Summa di Arithmetica*, which ended by stating that the solution of cubic equations was as impossible, given the current state of knowledge, as squaring the circle. This problem attracted the attention of mathematicians at the University of Bologna, who did not recognize negative numbers and so classified cubic equations into three types. (If negative numbers are admitted, this classification is unnecessary.) Scipio del Ferro, who died in 1526, solved all three types but failed to publish his methods.

Niccolo Fontana (*c.*1500–57 and nicknamed Tartaglia, 'the stammerer') rediscovered Del Ferro's results, and was eventually persuaded to reveal them to Hieronimo Cardano, 'the gambling scholar'. In 1545 Cardano published all these methods in his *Ars Magna*, together with the solution of fourth degree ('quartic') equations found by his student Ludovico Ferrari (1522–65). In 1572, following up certain aspects of these ideas, Raffael Bombelli (1526–73) became the first mathematician to suggest that square roots of negative numbers (now called 'imaginary') might have their uses in algebra.

FIBONACCI'S RABBITS

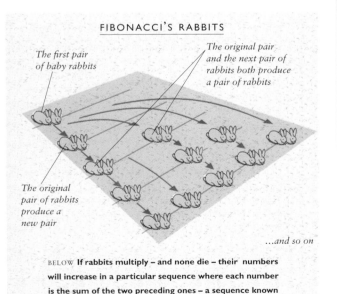

The first pair of baby rabbits

The original pair and the next pair of rabbits both produce a pair of rabbits

The original pair of rabbits produce a new pair

...and so on

BELOW **If rabbits multiply – and none die – their numbers will increase in a particular sequence where each number is the sum of the two preceding ones – a sequence known as Fibonacci numbers, from the nickname of the mathematician Leonardo of Pisa.**

EQUATIONS

An equation is an expression involving some unknown quantity **x**, which is set equal to some given value (usually 0). The aim is to find the value of x that 'solves' the equation – that is, renders it true. For example the equation $5x - 10 = 0$ has the solution $x = 2$. An equation like this one, in which x appears but higher powers x^2, x^3, and so on do not, is said to be *linear*. The general form is $ax + b = 0$, where a and b are specific numbers: in the example, $a = 5$ and $b = -10$. The solution is $x = -b/a$.

Next in order of complexity come quadratic equations which also involve the square of the unknown – for example, $x^2 - 5x + 6 = 0$. This equation has two solutions: $x = 2$, $x = 3$. The general form of a quadratic equation is $ax^2 + bx + c = 0$ and the solutions are $x = (-b \pm \sqrt{(b^2 - 4ac)})/2a$.

The symbol $\pm$ indicates that either the + or the - sign can be taken. When $b^2 - 4ac$ is greater than zero, this formula yields two distinct numbers. When $b^2 - 4ac$ is zero, both signs give the same answer (and there is only one solution). When $b^2 - 4ac$ is smaller than zero, the formula involves the square root of a negative number. At first such an expression was thought to be meaningless, but later a more extensive system of 'complex numbers' were invented, in which negative numbers possess square roots.

Next come cubic equations $ax^3 + bx^2 + cx + d = 0$ which usually have three solutions (though sometimes fewer), quartic equations $ax^4 + bx^3 + cx^2 + dx + e = 0$, with up to four solutions, and so on with quintic, sextic, septic, octic, and the like.

The Renaissance Italian mathematicians found formulas to solve cubic and quartic equations, but got stuck at the quintic. Later, Abel and Galois showed why: no such formula exists. But for several centuries, solving the quintic was a big open problem. Nobody worried about solving the sextic, though, because that would presumably be even harder than the quintic.

In general any expression of the above kinds, formed by adding together powers of x multiplied by various numbers, is called a polynomial, and the numbers are its coefficients. The highest power of x that occurs is called the degree. So, for example, $3x^8 + 2x^3 - 9x + 5$ is a polynomial of degree **8 (octic)** with coefficients **3, 2, - 9, 5**.

The search for formulas to solve polynomial equations should not be confused with the search for practical ways to solve them. Polynomial equations of high degree arise in practical questions such as calculating the APR, Annual Percentage Rate, on credit card bills, which can lead to an equation of the 23rd degree). Here what is needed is an approximation to the solution, not an exact formula. Many simple methods exist to find such approximations. In theoretical mathematics, however, approximations are less useful: the logic of proofs normally requires exact expressions.

HEAVEN AND EARTH

Big advances in mathematics have often gone hand in hand with big advances in cosmology. For some reason we often first notice a new mathematical pattern in the heavens, and only later apply it to problems on Earth. Just such an advance was made by Johannes Kepler (1571–1630), who became court astronomer to the Emperor Rudolph II (1552–1612).

Kepler looked for mathematical regularities everywhere from the planets to pomegranate seeds. In his day, there were thought to be just six planets: Mercury, Venus, Earth, Mars, Jupiter and Saturn. Why, Kepler asked, were there six planets? And what was the pattern underlying their distances from the Sun? He knew from the Ancient Greeks that there were just five regular solids. It occurred to him that between six planetary orbits, there are five gaps, so you can fit one of the five solids into each gap. His *Mysterium Cosmographicum* of 1596 presented an arrangement that gave pretty much the right sizes for the orbits, and it seemed as though he had hit on something. Unfortunately there are at least nine planets, as we now know. So Kepler was trying to explain something that wasn't true, and the fact that his calculations gave the correct results was pure coincidence. His investigations into other patterns, though, were to prove more useful.

By Kepler's time Polish astronomer Nicholas Copernicus (1473–1543) had discovered that the planets go round the Sun, although much of humanity, and the Roman church in particular, was somewhat resistant to this discovery. According to Copernicus the planetary orbits were almost circular, but not quite; Kepler wondered what the exact shape might be. He spent 20 years doing calculations for the planet Mars, and decided the orbit was an ellipse. He discovered

several other patterns: one relates the orbital period of a planet to the size of the ellipse, and another relates the planet's speed to its distance from the Sun. These patterns are known as Kepler's laws of planetary motion (*see Chapter 3*).

Kepler opened up the mathematics of the heavens. But when it comes to the mathematics of our Earthbound existence, we owe our greatest debt to Galileo Galilei (1564–1642), especially concerning the motion of bodies. Prior to Galileo, the great authority on motion was Aristotle (384–322 BC), whose view of 'mechanics' was based on human experience. For example, Aristotle theorized that a moving body eventually stops moving out of sheer exhaustion, and that cannonballs fall faster than feathers because they are heavier. Galileo realized that these effects are caused by friction and air resistance, and that the frictionless motion of bodies in a vacuum is a better context for an effective mathematical theory of motion. This led him to recognize that the key variable is not position or speed, but acceleration. He also realized that the path of any projectile subject to the downward pull of gravity, but encountering no air resistance, is a parabola. A parabola, as defined by the Greeks, being a cone sliced by a plane parallel to the edge of that cone. Or in other words, the trajectory of a cannonball, or a more modern tennis ball.

ABOVE **The ellipses, parabolas and other shapes a cone can be sliced into have provided the basis for all kinds of trajectories, from orbiting planets to spacecraft.**

RIGHT **Kepler's great discoveries about the orbits of the planets actually started from a misconception. He believed that the five gaps between the six planets then known arose because there are exactly five regular solids, each fitting into one of the gaps.**

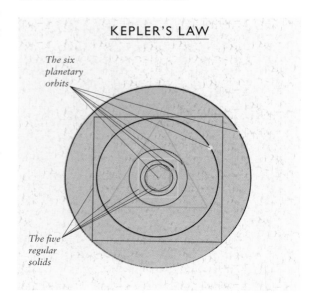

KEPLER'S LAW

The six planetary orbits

The five regular solids

THE FIRST UNIFICATION

The intellectual strength of mathematics derives in great part from its unity: every mathematical idea connects with, and influences, every other. The first big step towards today's integrated structure was the unification of algebra and geometry, which was carried out by French philosopher, René Descartes (1596–1650). To philosophers, Descartes' greatest intellectual contribution was his philosophy, but in fact few of his philosophical ideas about the workings of the world have survived the attentions of modern science. Mathematicians consider his mathematical work to be far more important, because it has more than just historical interest: its legacy remains relevant to this day.

Descartes' most influential discovery was that there is a deep connection between geometry and algebra. Consider the basis of a typical graph drawn on a sheet of paper: two lines, called axes, at right-angles to each other. The point where they cross is called the origin. The position of any point P can be given by measuring its distance from the origin along one axis, the x-axis, and along the other, the y-axis. The pair of numbers x,y is the coordinate representation of P. Descartes observed that geometric statements about such points – or sets of points, such as lines, circles and other curves – can be translated by a routine procedure into algebraic statements about the coordinates. For example, the geometric statement 'P lies on a circle of radius 1 centred at the origin' translates as 'the coordinates x,y of P satisfy the equation $x^2 + y^2 = 1$'. Conversely, any algebraic statement about x and y can be interpreted geometrically.

The importance of Descartes' idea – which amounted to a few casual remarks in a book about philosophy – is not that it renders either algebra or geometry obsolete. It is important because it allows both methods to be brought to bear on the same problem. Sometimes taking a geometric view leads to progress, and sometimes taking an algebraic view does. After Descartes, the mathematician could use whichever viewpoint was most fruitful, and swap between them at will.

CARTESIAN COORDINATES

Descartes discovered that both branches of mathematics – algebra and geometry – can be used to solve the same problem. So, for example, if a mathematician wants to find the coordinates of a point:

WORKING WITH COORDINATES

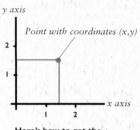

Here's how to get the coordinates of a point

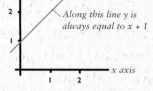

Here's a graph of those points (x,y) that satisfy the equation y = x + 1 (in this case a straight line)

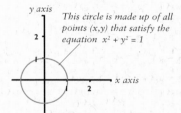

Here's a graph of those points (x,y) that satisfy the equation $x^2 + y^2 = 1$

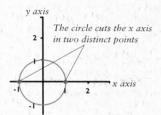

To show that the circle cuts the x axis at two distinct points we can use geometry as above or algebra as below

HOW TO WORK OUT COORDINATES USING ALGEBRA

Circle has equation $x^2 + y^2 = 1$
x axis has equation y = 0
Substitute second equation in first to get
$x^2 + 0^2 = 1$
i.e.
$x^2 = 1$
Solve to get
x = 1, x = -1
Recall that y = 0 so the solutions are the points with coordinates
(1, 0)   and   (-1, 0)
(These are, in fact, the coordinates of the two points marked with dots in the fourth diagram above.)

FATHER OF THE LAST THEOREM

Mathematics has a small number of notorious unsolved problems – problems asked many centuries ago, but which the concerted efforts of the world's mathematicians have been unable either to prove or disprove. Until 1994 a puzzle posed by Pierre de Fermat was somewhere in the top three or four. It took over 350 years for the world's mathematicians to find a solution.

Fermat was born in 1601. His father Dominique Fermat sold leather, and his mother Claire de Long was the daughter of a family of parliamentary lawyers. In 1648 Fermat became a King's Councillor in the local parliament of Toulouse, where he served for the rest of his life. He died in 1665, just two days after concluding a legal case. His profession was the law, but his passion was mathematics.

Fermat's most influential ideas concerned number theory: the study of ordinary whole numbers. Number theory was invented by Diophantus of Alexandria. We know very little about him: he was probably Greek, and if an ancient puzzle is to be believed he died aged 84. He flourished around AD 250, and he wrote a book on number theory called the *Arithmetica*. In it he gave, in particular, a complete answer to the problem of 'Pythagorean triples': find two squares that add up to another square, like $3^2 + 4^2 = 5^2$, or $5^2 + 12^2 = 13^2$.

Adding the square drawn on the side that is three units long...

RIGHT **Fermat did not have the space in the margin of** *Arithmetica* **to note the proof that he had discovered for Pythagorean triples. It took over 350 years to rediscover this proof.**

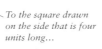

To the square drawn on the side that is four units long...

LEFT **Fermat's last theorem centres on what are known as Pythagorian triples. Pythagorus' famous theory about right-angled triangles shows that the squares of two sides of the triangle add up to the square of the side opposite the right angle, known as the hypotenuse. Pythagorian triples are the three sets of squares involved. Fermat's theorem asked whether cubes would add up in the same way as squares. Fermat contended that they would not.**

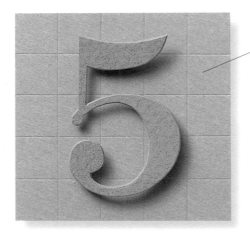

Equals the square drawn on the hypotenuse, which must be five units long.

PIERRE DE FERMAT
1601–65

Fermat was one of the most brilliant of all seventeenth century mathematicians and made many advances in number theory – the study of ordinary whole numbers. But he left a mysterious clue to an important proof known as Fermat's last theorem that intrigued mathematicians for centuries.

Fermat owned a copy of the *Arithmetica*, and he used to write down his ideas in the margin. Some time around 1637 he must have been thinking about Pythagorean triples, and he wondered what happens if, instead of squares, you try cubes. He got nowhere, and decided there must be a good reason for his failure. In his copy of the *Arithmetica* he scribbled the most famous marginal note in the history of mathematics:

> To resolve a cube into the sum of two cubes, a fourth power into two fourth powers, or in general any power higher than the second into two of the same kind, is impossible; of which fact I have found a remarkable proof. The margin is too small to contain it.

His 'remarkable proof' has never been found, and experts generally believe that, whatever he had in mind, it must have contained an error. His conjectured result became known as 'Fermat's last theorem' because it was the only theorem of his that nobody had managed to prove. In a sense it has done Fermat a disservice by diverting attention from his main contributions to mathematics, for those of his theorems that were proved changed number theory from a theoretical backwater into an essential part of the mathematical mainstream.

However, it should be remembered that Fermat did have other successes. For example, he was able to prove that a fourth power cannot be the sum of two fourth powers, so x (to the power of 4) + y (to the power of 4) = z (to the power of 4) will not yield a whole number answer for z.

LOGARITHMS

There is a great deal more to mathematics than just 'doing sums', or computation. Yet without major advances in computational technique mathematics would never have advanced to its current state, and it would never have held much attraction for the outside world. Aware of this, mathematicians have always demonstrated their skills and proved their worth by calculating things such as taxes, land, eclipses and horoscopes. Medieval and Renaissance mathematicians even engaged each other in public tournaments to establish who was Top Gun.

Until the development of practical electronic computers in the middle of the twentieth century, one improvement in computational technique stood out above all others: the invention of logarithms. Using logarithms replaces the complicated problem of multiplying two numbers together by the far simpler task of adding two related, but different numbers. And exceedingly tricky operations, such as extracting the seventh root of a number, are likewise turned into entirely tractable problems. Logarithms gave mathematics a major boost by making it a practical proposition for a a huge range of applications.

Logarithms were invented in about 1594 by John Napier, a Scot, who published the idea in 1614. He came up with the idea in the context of spherical trigonometry, as used in astronomy, map-making and navigation, so the actual form in which it appeared concealed the true simplicity of the concept. New mathematical ideas are often first recognized in a complicated form and later distilled down to their essence, and Napier's approach was substantially improved and clarified by Henry Briggs in 1615.

Using logarithms involves acquiring a book of tables, in which the logarithm of each number is computed once and for all. To multiply two numbers, you simply find their logarithms in the table, add them to come up with another logarithm, and then consult the table to find the number this represents. For example, the calculation $48.73 \times 678.34 = 33{,}055.51$ is easy enough if you have a calculator, but otherwise tedious. Finding the logarithms of the numbers makes it a simple matter of addition:

logarithm 48.73: 1.6877964
logarithm 678.34: 2.8314474
add them together: 4.5192438

Looking up 4.5192438 in the table reveals that it is the logarithm of 33,055.51.

Although logarithms are no longer used for multiplication, having been superseded by calculators and computers, they remain of vital importance because of the way they relate multiplication to addition.

**WILHELM
GOTTFRIED LEIBNIZ**

1646–1716

The great German
mathematician Leibniz was
one of the two originators of
the branch of mathematics
called calculus, which centres
on rates of change.

BELOW **With the invention of
calculus mathematicians had a
wonderful tool for calculating
things such as how a cannonball
flies through the air. But
seventeenth-century
mathematicians fought tooth
and nail over who invented it.
Was it Newton, who had the
idea first? Or Leibniz, who
published first?**

THE NEWTONIAN REVOLUTION

Isaac Newton, born in Lincolnshire, England, in 1642, is a pivotal figure in both mathematics and physics. Yet as John Maynard Keynes perceptively noted, he was not so much the first of the moderns as the last of the magicians. There was a mystical side to Newton; he had an extensive interest in alchemy, for example. He showed little sign of genius until, forced to leave Cambridge for his family home in 1665–66 to avoid the plague, he occupied his time by working out many of the key ideas that later appeared in his masterpiece, the *Principia Mathematica Philosophiae Naturalis*.

In this three-volume work Newton laid down his 'system of the world'. Taking hints from Kepler, Galileo and others, he stated mathematical laws for moving bodies and for gravitational forces. Armed with rudimentary but effective mathematical techniques, he applied his new insights to physics. There he derived the elliptical form of planetary orbits, the general principles of wave motion in air (as sound) and in water, the mass of the Sun, the density of the Earth, the fact that the Earth is flattened at the poles, the precession of the equinoxes, the main features of tides, certain irregularities in the motion of the Moon, and the paths of comets. Any one of these achievements would have merited the Nobel prize, had it existed in Newton's day.

Yet many of Newton's mathematical achievements are perhaps even more significant than his physics. Unpublished documents – the 'Portsmouth papers' – show that at the time of writing the *Principia*, Newton had the calculus at his fingertips. Calculus is the mathematics of instantaneous rates of change, and it has two main branches. Differential calculus permits the calculation of these rates, with geometric applications such as finding tangents to curves. Integral calculus does the converse: given the rate of change of some quantity, it derives the quantity itself. The more direct geometric applications of integral calculus are to the computation of areas and volumes. Newton didn't develop calculus to answer his questions about gravity and falling bodies, but to help him in his research.

Newton did not present the material in the *Principia* using calculus; instead, he decided to cast his arguments in classical geometric form. The result is a geometric *tour de force*, all the more impressive because the intuition behind it is actually concealed. On reading it, you think, 'How could Newton possibly have thought of such strange geometric proofs?' The answer is, 'Because he derived them from ideas that he kept up his sleeve.' Newton's version of calculus was published only after his death, as his *Method of Fluxions* of 1732.

The same ideas, expressed rather differently, formed a major part of the life's work of the German mathematician and philosopher, Wilhelm Gottfried Leibniz (1646–1716), and this dual discovery led to a heated but pointless controversy between the English mathematicians and their continental counterparts. The truth is simple: Newton had the ideas first, but Leibniz – without knowledge of Newton's work – published them first. Neither stole anything from the other.

From the modern point of view, Newton's chief achievement was not calculus as such, but the sweeping realization that nature can be modelled using differential equations – equations that involve rates of change. For example, his main law of motion states that the acceleration of a body is proportional to the force acting upon it. Acceleration is the rate of change of velocity (with respect to time), and velocity is the rate of change of position (again with respect to time). So Newton's law states a property of *a rate of change of a rate of change* of the quantity that is most readily observed – position.

However, the law for acceleration is simple and general, whereas any attempt to formulate a law of motion using position falls apart because of the wide variety of *changing* forces that may act.

The simplicity of the mathematical law for acceleration leads to a gut feeling that acceleration is primary while position is secondary, and the philosophical view that nature is, at root, simple. If we only look at nature in the right way – Newtons's way – a wonderful and satisfying order emerges from apparent chaos.

DIVIDING ZERO BY ZERO

Calculus is not about numbers, but functions: mathematical rules, often presented using formulas, that assign to each number x in some range another number f(x). For example, the 'square' function is defined by the rule $f(x) = x^2$. The key idea of calculus is the derivative, or rate of change, of a function. Both Newton and Leibniz in effect defined it as follows: change the value of the variable x by a very small amount, and divide the change in the value of the function by that same small amount. Then let the change become vanishingly small and see what you get.

This kind of caclulation was attacked by Bishop Berkeley, the Irish churchman, philosopher and mathematician, on the grounds that it amounts to dividing zero by zero and therefore makes no logical sense. The mathematicians sought refuge in physical analogies and the 'spirit of finesse' as opposed to the 'spirit of logic', but Berkeley had them floundering. It took over a century to answer his objections, with a careful definition of 'passing to a limit'; calculus thereby became a logically deeper subject, called analysis. Nobody except Berkeley worried much, and calculus flourished despite this logical flaw.

THE FUNDAMENTAL LAWS

The whole of algebra rests on certain general properties of numbers, and these properties have technical names. Suppose that a, b, c are numbers. Then the main properties are:

Commutative law for addition – numbers add up to the same total no matter what order you add them:

$$a + b = b + a$$

Associative law for addition – numbers add up to the same total no matter how you group them:

$$a + (b + c) = (a + b) + c$$

Commutative law for multiplication – numbers multiply together to the same total no matter what order you multiply them:

$$ab = ba$$

Associative law for multiplication – numbers multiply together to the same total no matter how you group them:

$$a(bc) = (ab)c$$

Distributive law – when numbers are both added and multiplied, the way they are grouped affects the total:

$$a(b + c) = ab + ac$$

NEWTON'S LEGACY

Newton's successors were quick to exploit his breakthrough. One of the most active was Leonhard Euler, born in Basle, Switzerland in 1707, and surely the most prolific mathematician in history. Euler's career is intertwined with the conquest of the seas, perhaps the major challenge of his period. The use of celestial bodies in navigation in vogue at that time spurred Euler to develop the first calculable answer to the motion of three bodies – the Sun, Moon and Earth.

Euler ranged over the whole of mathematics, from number theory to mechanics. He wrote innumerable elegant textbooks, which not only systematized the work of his time but extended it. His 1736 book *Mechanics* was especially influential: it recast the geometrical viewpoint of Newton's *Principia* into explicit analysis, bringing to bear the full power of calculus. His approach to a rigid body moving in space is still used today, remembered in the term 'Euler angles'. He wrote three major works on analysis itself but his first major impact on mathematics, was made in the calculus of variations: the 1744 *Methodus Inveniendi Lineas Curavas Maximi Minimive Proprietate Gaudentes* (Methods of Investigating Longest and Shortest Properties of Curved Lines).

As the title suggests, the aim of the calculus of variations is to find curves or surfaces that maximize or minimize some property. The calculus of variations originated in the problem of the 'brachistochrone' – if you roll a ball down a curve, what shape of curve allows the ball to reach the bottom fastest? Johann Bernoulli solved this problem in 1696: the answer is an inverted cycloid – the path formed by a point on a circle as the circle rolls along a straight line. Euler also made inroads into mathematical physics, especially hydrodynamics and elasticity, and his fingerprints are all over today's mathematics.

THE WAVE EQUATION

Newton's methods for solving differential equations were rudimentary – a fact that did not prevent him making excellent use of them, so free-ranging was his talent. His successors extended the ideas of differential equations into every area of mathematical physics – fluid dynamics, heat, electrostatics, magnetism, sound, light – by extending the concept to 'partial differential equations', which involve rates of change with respect to several different variables, such as space and time.

One of the most important partial differential equations is the wave equation, which describes the shape and movement of waves of all kinds. The wave equation emerged from early work on the vibrations of a violin string, although it had no practical application to the design of musical instruments in those days, and was regarded as a purely intellectual problem. Now that computers are able to solve both the basic wave equation and the less idealized variants for the complicated shapes and materials of real instruments, such applications are routine. Probably the earliest big result was obtained in 1714 by the English mathematican, Brook Taylor (1685–1731), who calculated the fundamental vibrational frequency of a violin string in terms of its length, tension, and density. The string can produce many different musical notes, depending on the position of the 'nodes', or rest-points. For the 'fundamental' frequency, only the end points are at rest. If the string has a node at its centre, then it produces a note one octave higher. The more nodes there are, the higher the frequency of the note, and these higher frequencies are called overtones.

The vibrations are shown as standing waves – the shape of the string at any instant is the same, except that it is stretched or compressed in the direction at right angles to its length – and the

LEONHARD EULER
1707–83
Euler was one of the most prolific of all mathematicians, making important discoveries in almost every field of mathematics including solving the formidable three-body problem – the relative movement of the Earth, Moon and the Sun in motion. Euler set calculus and trigonometry in their modern forms, creating differential calculus virtually single-handedly.

ABOVE LEFT **One of the problems that Euler focused on was the 'brachistochrone' – that is the shape of the curve for a slope which allows a ball to roll down fastest. The problem was solved later by Euler's compatriot, Johann Bernoulli.**

ABOVE **Euler made fundamental contributions to the mathematics of waves by studying the frequency of vibrations in both drums and violin strings. Euler's discoveries provided the basis for understanding electromagnetic waves – and so underpin all modern telecommunications.**

more interesting boundary. Boundaries are absolutely crucial to this whole subject. The boundary of a drum can be any closed curve, and the key condition is that the rim of the drum remains fixed. This 'boundary condition' greatly restricts the possible motion of the drum.

The problems posed by violin strings and drums may seem limited in their relevance, but the mathematicians of the eighteenth century extended the domain of the wave equation to many areas of physics, revealing just how important it is. Waves arise not only in musical instruments, but also in the physics of light and sound. Euler found a three-dimensional version of the wave equation which he applied to sound waves. Roughly a century later, James Clerk Maxwell (1831–79) extracted the same mathematical expression from his equations for electromagetism, and predicted the existence of radio waves. Without the early mathematicians' work on musical instruments, we would not have television.

ABOVE **In 1748 Euler worked out the equation for the wave representing the vibrations of a violin string.**

waveforms are sinusoidal in shape. But in 1746 Jean le Rond d'Alembert showed that the full story isn't quite that simple. There are many vibrations of a violin string that are not sinusoidal standing waves. In fact he proved that the shape of the wave can start out being anything you like. Euler explored the question much further, and by 1748 he had worked out – and solved – the wave equation for a string. These discoveries started a century-long controversy, ending with the result that you can actually get all possible vibrations of the string merely by superposing these sinusoidal waves in suitable proportions.

In 1759 Euler extended the work to drums, deriving a wave equation for a two-dimensional membrane. Drums differ from violin strings not only in their dimensionality – a drum is a flat two-dimensional membrane – but in having a much

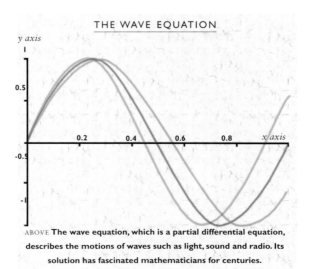

THE WAVE EQUATION

ABOVE **The wave equation, which is a partial differential equation, describes the motions of waves such as light, sound and radio. Its solution has fascinated mathematicians for centuries.**

HEAT, MECHANICS
AND THE SOLAR SYSTEM

Another key partial differential equation is the heat equation, introduced by Joseph Fourier (1768–1830) in 1822 in his *Théorie Analytique de la Chaleur* on the physics of heat flow. For example, imagine taking a metal rod and heating one end. What will the temperature be at the other end, two minutes later? What will the temperature be at any chosen point along the rod? The heat equation determines the answer, but not directly: instead it relates how the temperature changes in space (along the rod) to how it changes in time. To find the actual temperature, you have to solve the heat equation.

Fourier worked out a wonderful method to do this, but introduced his idea in an amazingly clumsy manner. He claimed to have proved that any solution can be written as an infinite series, whose terms were expressions in the sine and cosine function from trigonometry. For example, the solution might look like this:

$$\sin(x) + \tfrac{1}{2}\sin(2x) + \tfrac{1}{3}\sin 3x + \ldots - \cos(x) - \tfrac{1}{4}\cos(2x) - \tfrac{1}{9}(\cos 3x) - \ldots$$

Here the numbers $1, \tfrac{1}{2}, \tfrac{1}{3}\ldots$ and $-1, -\tfrac{1}{4}, -\tfrac{1}{9}\ldots$ are called coefficients. To solve the heat equation under given conditions, all you have to do is find the coefficients in this 'trigonometric series', and Fourier had a method for doing that.

The method seemed to work, in practice, but it wasn't at all clear that it was based on sound logical principles. In particular, it wasn't clear that this trigonometric series made sense, or that every solution would possess such a series. Fourier claimed to have a 'proof', but it rambled all over the place to get to its conclusion. The conclusion was nice, though: an explicit formula for the coefficients of the series in terms of certain integrals, also involving trigonometric functions sin and cos. By the time of Dirichlet and Riemann much simpler ways had been found to derive Fourier's formula for coefficients. The scene was set for a major crisis of confidence, half a century later, when mathematicians finally realised just how tricky the whole area was. It took another fifty years or more before the problems were sorted out, the conclusion being that Fourier's formula was correct provided certain technical conditions were valid. The important new discovery was what those conditions were. In other words, Fourier's 'proof' was nonsense, but his answer was often right: after a century's hard work, everyone finally knew exactly when that was the case.

LAGRANGE AND LAPLACE

Euler may have been the most prolific mathematician of all time, but in terms of originality he had to take second place to the French mathematican Joseph-Louis Lagrange (1736–1813), who is now generally considered the greatest mathematician of the eighteenth century. Lagrange was born into a rich family in 1736, but his father lost the family fortune in a series of doomed speculative business ventures. Mathematics was the beneficiary of Lagrange's misfortune, for he had to find some way of earning a living.

Lagrange took to the subject almost by accident, having come across an essay by Edmund Halley (1656–1742) (best known for his work on the comet named after him) on the virtues of Newtonian calculus. Lagrange's talent was prodigious: by the age of 16 he was Professor of Mathematics at the Royal Artillery School in Turin, and before he was 23 he had conceived – and must also have explored in some depth – his master project: to use the calculus of variations as a foundation for the mechanics of both solids and fluids. This project occupied much of his working life: the *Mécanique Analytique* was finally published in 1788 when he was 52.

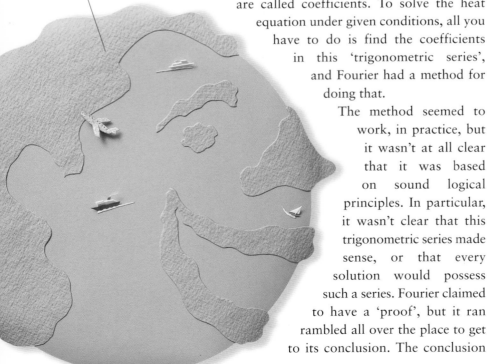

BELOW **Normally, we always see the same face of the Moon from Earth because, although the Moon is always slowly turning, it is also travelling slowly round the Earth. The result is that half of the Moon remains forever out of sight.**

The Earth's gravity controls the turning of the Moon

Meanwhile Lagrange worked on number theory, probability, sound and the vexed question of the vibrating string. In 1764 he solved, the very difficult problem of the libration of the Moon. The Moon always presents the same face to the Earth, as a result of tidal friction, but there are slight irregularities in its position: the most significant of these is known as 'libration'. He also greatly advanced our understanding of the motion of Jupiter's moons – a six-body problem, when the important influences of Jupiter and the Sun are taken into account.

The career of another French mathematician, Pierre-Simon Marquis de Laplace (1749–1827), has much in common with Lagrange's. In his day, Laplace was generally considered a better mathematician than Lagrange, but the verdict of posterity reverses the order. Laplace's greatest work was in celestial mechanics, and he set his sights on the big question: the motion of the entire Solar System.

The high point of his work is his *Mécanique Céleste*, published in parts between 1799 and 1825. In this work he laid the basis for the whole subject, making major inroads into questions such as the stability of the Solar System. Will it continue indefinitely in its present form, or might a planet be lost to the outer reaches, or crash into the Sun? Today we know that, thanks to the chaotic properties of nonlinear dynamical systems, a definite answer to such a question cannot be given. The motion is too sensitive to the precise state of the Solar System at any given instant: stable states and unstable ones are densely intertwined. But without Laplace's seminal contributions, we would never have reached the stage at which we could justify such an answer.

THE FOX OF MATHEMATICS

Arguably the greatest mathematician whoever lived was Carl Friedrich Gauss, born in 1777. Gauss, the son of a farm hand, was precocious – it is said that he corrected his father's arithmetic at the age of three. By the time he was 14 his fame had spread, and Carl Wilhelm Ferdinand, Duke of Brunswick, began to support his education. At the age of 19 Gauss was trying to decide whether his future lay in linguistics or in mathematics when he made an epic discovery.

The Greeks had known how to use a ruler and compasses to construct regular polygons with 3, 5, and 15 sides, and how to double those numbers as many times as required. Everyone assumed that no other polygons could be constructed in this way, but on 30 March 1796 Gauss discovered how to construct a regular 17-sided polygon. This unexpected result led Gauss to plump for mathematics, and he began to keep a cryptic mathematical diary in which he recorded his big discoveries. The construction of the regular 17-gon is the first of its many entries. (Later mathematicians used Gauss' methods to construct regular polygons with 257 and 65,537 sides – the only known new possibilities.)

Gauss' great love was number theory – the study of whole numbers. It might appear that such lowly materials were unworthy of the attention of the greatest mathematician of all time, but exactly the contrary is true. Whole numbers are wonderfully subtle, and many unsolved problems in mathematics relate to them – Fermat's last theorem, discussed earlier, is just one example. Simplicity of concept often goes hand in hand with intellectual depth, and so it is here. Gauss' work in number theory is laid down in his massive *Disquisitiones Arithmeticae*, whose crowning glory, the 'law of quadratic reciprocity', caused even the great Gauss many years of hard thought.

Early in Gauss' notebooks we find the entry EYPHKA! num = $\Delta + \Delta + \Delta$ (Eureka! Number = triangular + triangular + triangular). Recall that

The Moon tends to show the same face to the Earth

ABOVE **Although the Moon normally shows the same face to the Earth, it has a slight wobble in its orbit known as libration. This irregularity means that every now and then it turns slightly to each side alternately, revealing just a little bit more of the hidden side. In 1764 the French mathematician Joseph Louis Lagrange worked out just why.**

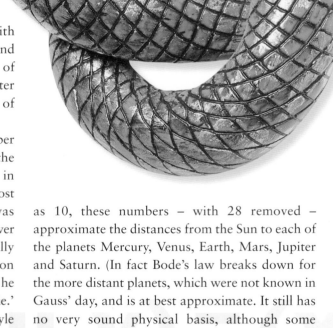

A rubber ring may look more like a ball than a coiled snake does – but topologically it is only the snake and ball that are equivalent, because the coil can be deformed (theoretically) into a ball; the rubber ring cannot

the triangular numbers 1, 3, 6, 10, 15 and so on fascinated the Pythagoreans. Gauss proved that every number is a sum of three triangular numbers – for example, 28 = 3 + 10 + 15. A mathematically equivalent statement is that every number of the form 8n+3 is the sum of three odd squares – a deeper refinement of Lagrange's already deep result that every whole number is a sum of four squares.

Another of Gauss' triumphs was a full understanding of complex numbers: numbers of the form x+iy where i = √−1. He proved that every polynomial equation with complex coefficients has a complex solution, and he wrote down the beginnings of a theory of analysis for complex-valued functions. This later flowered into full bloom with the work of Augustin-Louis Cauchy, published in 1825.

Gauss' achievements went well beyond number theory. In one way or another he laid the foundations for much of what happened in mathematics in the next century. Many of his most radical ideas went unpublished – he was obsessively anxious about public criticism by lower intellects. Even when he did publish, he carefully polished his ideas until only a beautiful skeleton remained. 'When a fine building is constructed,' he said, 'its scaffolding should no longer be visible.' But other mathematicians found this austere style unhelpful: Carl-Gustav Jacobi complained that Gauss was 'like the fox, who erases his tracks in the sand with his tail'.

STAR QUALITY

At the age of 24, Gauss' attention was drawn to the applications of mathematics by the discovery of a new 'minor planet' or asteroid named Ceres, which filled a gap between Mars and Jupiter in a curious empirical pattern in planetary distances known as Bode's law.

Take the series 0, 3, 6, 12, 24, 48, 96, doubling at each step after the first, and add 4 to get 4, 7, 10, 16, 28, 52, 100. Taking the Earth's orbital radius as 10, these numbers – with 28 removed – approximate the distances from the Sun to each of the planets Mercury, Venus, Earth, Mars, Jupiter and Saturn. (In fact Bode's law breaks down for the more distant planets, which were not known in Gauss' day, and is at best approximate. It still has no very sound physical basis, although some theories of planetary formation lead to similar distributions of distances.)

What body, then, lies at distance 28 in Bode's scheme? The answer is the asteroid belt and the minor planet Ceres; found on the first day of the nineteenth century by Giuseppe Piazzi. This curiosity made Gauss a scientific superstar, for Ceres disappeared behind the Sun's light soon after its discovery, and was in danger of being 'lost' to astronomy. Gauss rose to the technically difficult challenge of predicting where it would reappear, from inadequate observational data. To do so, he introduced a whole range of new computational methods into celestial mechanics, including the

ABOVE **One of the most surprising and fruitful areas of mathematical advance over the last century has been topology – the study of surfaces as they are twisted or stretched, first hinted at in unpublished papers on knots by Gauss. It is sometimes known as 'rubber sheet' geometry. New shapes that can be made simply by twisting, bending or stretching are topologically equivalent; shapes that can only be made by cutting, tearing or folding are not.**

'method of least squares' to fit a straight line to irregular data.

Inspired by his success, Gauss spent the next 20 years calculating orbits: the orbits of comets and the orbits of Pallas, Vesta and Juno – Ceres' cosmic sisters. Some mathematical historians consider this to have been a tragedy for mathematics, but even the greatest minds sometimes need to seek new intellectual hunting-grounds.

Gauss also made inroads into surveying, based on his extensive understanding of the geometry of surfaces. He investigated electricity and magnetism, taking some important steps along the path that led, via Michael Faraday, to James Clerk Maxwell and a complete mathematical theory of electromagnetism. Together with Wilhelm Weber, Gauss invented the first practical telegraph. And in Gauss' unpublished papers, there are tentative steps towards a theory of knots and surfaces that later flowered into the centrepiece of twentieth century mathematics: topology, the maths of surface properties that do not change under distortion.

BELOW **Karl Friedrich Gauss (1777–1855)**
was perhaps the greatest of all mathematicians,
making important discoveries in virtually every
field of mathematics, from number theory
to the orbits of asteroids.

REFINED BUT USELESS?

So-called 'imaginary' numbers – square roots of negative quantities – first showed up in **Cardano's** *Ars Magna* of 1545, where they were dismissed as being 'refined but useless'. Soon afterwards, **Raffael Bombelli** discovered that manipulations with imaginaries illuminated some tricky problems in the solution of the cubic equation, converting apparently meaningless formulas into genuine solutions. Later mathematicians like **Descartes** and **Newton** saw the occurrence of imaginaries as evidence that a problem was insoluble – and so showed that they hadn't fully taken Bombelli's ideas on board. By the time of **Euler**, imaginary numbers had become an established part of mathematical technique, but their meaning remained mysterious.

Euler introduced the symbol $i = \sqrt{-1}$ and happily manipulated 'complex' numbers (this term merely signifies 'twofold' and does not indicate that complex numbers are actually complicated). By 1702 the rudiments of complex calculus existed, and **Johann Bernoulli** calculated integrals using logarithms of complex numbers. No-one could define just what complex numbers really are, but in time mathematicians put this vexed question aside to work on how they could be used to solve major problems in analysis and applied mathematics.

In 1811, **Gauss** proved a key property of complex integrals – that the integrals of a complex function along two paths in the complex plane, with common end-points, are equal unless the region bounded by the curves contains a 'singularity' where the function fails to be defined or is defined but fails to be differentiable. This result is now known as 'Cauchy's theorem', because **Augustin-Louis Cauchy** was the first to publish it, in 1825. Cauchy built on this insight to develop an astonishingly useful theory of complex analysis. Mathematicians embraced **Cauchy's** ideas in droves, and the awkward question, 'What are complex numbers?' was ignored.

Cauchy slid over the difficulties of concepts such as 'limit' and 'convergence' for infinite series. In 1843 **Pierre-Alphonse Laurent** introduced 'Laurent series': expansion of complex functions involving both positive and negative powers of the variable. Laurent's approach was seized upon by **Karl Weierstrass**, who based the whole of complex analysis on Laurent series. Only after Weierstrass was analysis truly rigorous: his 'epsilon-delta' definitions still form the basis of undergraduate analysis courses, baffling students as much as they initially baffled **Weierstrass's** contemporaries.

From the mid-nineteenth century onwards the progress of complex analysis was rapid and extensive. Complex numbers, once decried as useless, found a lasting place in mainstream mathematics, with fundamental applications to aerodynamics, fluid mechanics, electrical engineering and electronics. They also proved indispensable to wave mechanics and quantum theory. Useless indeed…

ÉVARISTE GALOIS
1811–32

Galois' ideas, discovered after his premature death, became the basis of what is now known as group theory – a fundamental concept in modern mathematics.

BELOW **Legend has it that Galois was killed at the age of just 20 in a duel over a woman which became known as the 'infamous coquette'. However, it seems more likely that the duel was fought with a political opponent, angered by his tirades against the French King Charles X.**

THE INFAMOUS COQUETTE

One of the great classic problems, running like a thread through centuries of mathematics, is the solution of the quintic (fifth degree) equation. We've already seen how the Renaissance Italian mathematicians discovered solutions to cubic and quartic equations. The solution of quadratic equations was no secret to the ancient Babylonians, around 1600 BC. In each case the solutions are expressed in terms of radicals – algebraic formulas in the coefficients, involving nothing more complicated than nth roots for various n. Can the quintic also be solved by radicals? Nobody knew.

Euler tried and failed, but came up with a new method for the quartic. Lagrange unified all of the tricks used by his predecessors, relating them to 'permutations' of the equation's roots – ways of writing them in order. Given three roots x, y , z, say, there are six permutations: xyz, xzy, yxz, yzx, zxy and zyx. Lagrange showed that the unified trick failed on the quintic. From that time on there was a general feeling in the air that radicals would not suffice to solve quintics, and the first attempted proof of this was given in 1813 by Paolo Ruffini. His proof was published in an obscure journal, and in any case it had gaps; the first accurate proof was given by Neils Hendrik Abel in 1824.

In 1832 a young Frenchman, Évariste Galois, was killed in a duel, ostensibly over a woman ('an infamous coquette', Galois wrote at the time, but she seems to have been Stéphanie du Motel, the entirely respectable daughter of a doctor living just down the street from Galois). Galois had a fiery temper and an instant dislike for anyone he considered his intellectual inferior. He had for some time been trying to obtain official approval of his mathematical theories, but his personality didn't improve his prospects. Only in 1843 did Joseph Liouville understand what Galois was on about, when he announced some of the dead duellist's discoveries to his peers.

Galois had picked up on Lagrange's ideas about permutations, and discovered that these held the key to whether an equation was soluble by radicals. The secret was the equation's 'Galois group': those permutations of its roots that leave unchanged any algebraic relation between them. In modern terms, the Galois group is the set of all algebraic symmetries of the equation. Galois proved the insolubility of the quintic by radicals.

The Galois group marked a turning point in algebra. Before Galois, algebra was a symbolic system for establishing general truths about numbers. After Galois, it was a symbolic system for establishing general truths about anything. The raw materials might be numbers – but they might be permutations, groups of permutations, or other more abstract entities. All that mattered was to set up a consistent symbolism.

Perhaps the most important abstract algebraic concept to emerge from this viewpoint was the concept of a group. This is a collection of mathematical objects that can be combined in pairs to create a new object, also in the collection – subject only to a few simple algebraic rules, the most important being the 'associative law' x(yz) = (xy)z. Today's mathematics relies heavily on groups, in part because group theory provides a systematic method for exploring and exploiting the symmetries of mathematical objects and processes. Symmetry runs deep in mathematics, and a great deal relies on it.

INSIGHT AND HYPOTHESIS

Some mathematicians are prolific; Euler is the extreme example. Others make an equal impact by creating a small body of work, but one of unsurpassed intellectual depth and insight. Such a mathematician was Georg Bernhard Riemann (1826–66). In order to qualify for the position of *Privatdozent* in nineteenth-century Germany – roughly speaking, a lecturer who could charge pupils for his services, the first serious foothold on the academic career ladder – the candidate had to propose a list of topics to speak on. Gauss, his examiner, picked the most difficult: the foundations of geometry.

Riemann responded by taking the various types of non-Euclidean geometry then known, and viewing them as special cases of a far more general kind of geometry, the geometry of geodesics in curved spaces. A geodesic is the shortest path between two points. In the plane, geodesics are straight lines. In spherical geometry, they are great circles. In the types of non-Euclidean geometry known in Riemann's day, they are the non-Euclidean analogues of straight lines. But in all cases, the space in which those lines or curves existed had a simple, uniform geometry.

Not so for Riemann. To him, space could be whatever you wanted it to be. Even its dimensions were up for grabs: not just the classical two or three dimensions, but four, five, a million, whatever. And the geometric properties of the space could vary from place to place. All you needed was a manifold (a generalized system of coordinates) and a metric (a notion of distance that applied, in the first instance, only to nearby points). From the metric, you could construct short geodesics; then by stringing these together over longer distances you could set up large-scale notions of distance and geometry. Most importantly, you could compute from the metric the extent to which the space was curved, and whether the curvature was positive (like a bowl) or negative (like a horse's saddle).

Gauss was known to be fascinated by curvature and intrigued by non-Euclidean geometry, so it's hard not to conclude that Riemann was an adept academic politician as well as a consummate mathematician. At any rate, his reputation was made. Today, Riemann's geometry lies at the heart of Einstein's theory of general relativity.

Riemann's other great contribution was to number theory. He discovered deep connections between the theory of prime numbers and the mathematics of complex functions. In a sense, the idea is that a complex function can 'encode' the properties of an entire sequence of numbers. When this sequence is the prime numbers, the resulting function – which Riemann denoted by the Greek letter 'zeta' and is therefore now called the zeta-function – turns out to be extremely interesting from the viewpoint of complex analysis.

The kind of theorem that can be proved using Riemann's insights into the connection between integers and analysis is exemplified by the prime number theorem: the number of primes less than a given number x is asymptotic to $x/\log x$. (Asymptotic means approximately equal to, with the approximation getting better and better as x gets larger.) In this manner Riemann created an entirely new branch of mathematics called analytic number theory. This theory has no application, it is an entirely intellectual pursuit.

He also left unresolved what is probably the greatest open problem in mathematics: the Riemann hypothesis. This has to do with the location of the zeros of Riemann's celebrated zeta-function: apart from certain trivial exceptions, they should all lie on a single line in the complex plane. All the evidence is that the Riemann hypothesis is true – and valid new techniques such as a prediction of error in the likely rate of error, have been motivated merely by assuming that it is, and seeing where the assumption leads – but the proof remains elusive. In fact nobody knows how to start.

GEORG BERNHARD RIEMANN

1826–66

German mathematician Riemann produced few papers in his life, but they were 'perfectly formed', opening an entirely new form of geometry relating to curved surfaces.

THE SIZES OF INFINITY

Georg Cantor (1845–1918) is a wild card in the mathematical brat pack. He was a prime mover in the push towards increased abstraction in mathematics, a trend that by the 1960s had led many outside observers to believe that the whole subject was becoming totally divorced from anything of practical value. Oddly, nothing could have been more false. Abstraction is the mathematician's way of keeping the subject under control, for abstraction goes hand in hand with generality. When the same basic idea comes up in 17 different areas of mathematics, it is either necessary to descibe each instance in detail, or have the whole lot subsumed into something more concise – but one step further removed from reality than each of those 17 instances.

Is mathematics discovered or invented? Probably neither, for mathematics is a virtual subject, the 'material' of which consist of ideas. However, we can make a useful distinction between mathematics that explores new aspects of existing traditions (discovery), and makes radically new departures (invention). In some sense, discoveries are sitting around in conceptual space waiting to be made, whereas inventions expand the whole

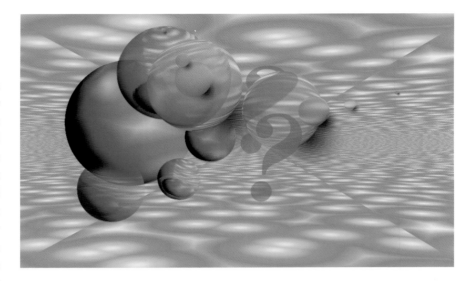

space of ideas into new dimensions. Discoveries will eventually be made by somebody; inventions, if not made, may die with their would-be inventor. Cantor, in this sense, was an inventor, and his invention was set theory.

Ironically, Cantor's radical expansion of the conceptual space in which mathematics flourishes arose from an entirely conventional investigation into a deep-seated difficulty in analysis. For half a century, Fourier's theory of trigonometric series had been causing a growing crisis, in which different theorems, apparently proved in all customary rigour, contradicted each other. The crisis was eventually resolved by raising the standard of logical rigour considerably, and Cantor was a prime mover in this development. It remains one of the best reasons for not relaxing the requirement of rigorous proof in mathematics, although innumerable well-meaning but ill-advised people continue to call for just such an abandonment of responsibility on the grounds that it makes mathematics easier. (It does, just as abandoning the driving test makes learning to drive easier.)

Cantor's work on Fourier series led him to focus on sets – or collections – of points on the real number line. These might be the set of points at which a Fourier series misbehaves in some manner: it fails to converge, or converges to the wrong sum, for instance. The subtleties of Fourier analysis hinged on abstract, general properties of such sets – another step towards topology, but also

INFINITE SETS

0	1	2	3	4	5
0	1	-1	2	-2	3
6	7	8	9	10	...
-3	4	-4	5	-5	...

ABOVE **The study of different sets, not only sets of numbers, led Georg Cantor to the realization that there are many different sizes of infinity. The most familiar infinite set is ordinary 'natural' numbers, the numbers we use every day. These numbers can be matched with negative numbers to create an infinity twice as big.**

the seeds of revolution. For what really mattered seemed to be abstract properties of sets of any kind, not just sets of numbers. And this led Cantor to an epic realization – that there can be different sizes of infinity.

The most obvious infinite set is the set {0, 1, 2, 3...} of natural numbers – numbers that are ordinary integers, used every day. (Curly brackets { } are used by set theorists to enclose a list of what is in the set.) Cantor observed that apparently 'bigger' sets, such as the integers, positive *and* negative, could be matched one-to-one with the natural numbers: that is, the two sets have the same 'cardinality'.

All very well: we know that the infinite behaves in curious ways, and in a sense Cantor's observation just says that $\infty + \infty = \infty$, using '∞' to represent 'infinity' – an equation that raises few eyebrows in mathematical circles. However, Cantor went on to show that the set of all real numbers, all infinite decimals, *cannot* be matched one-to-one to the natural numbers. The natural numbers are 'countably' infinite, the real numbers 'uncountably' infinite. A bigger size of infinity. More than that: given any infinite set, there is another 'bigger' one that cannot be matched to it. The range of different infinities is itself infinite.

BELOW **Möbius bands are made by twisting a strip of paper through 180° and joining the ends, converting a two-sided piece of paper into a shape with just one continuous surface.**

A line can be drawn down the middle of both sides of the strip without lifting the pen from the paper

THE GEOMETRIZATION OF MATHEMATICS

Real mathematics, complete with logical structure and a notion of proof, began with the geometry of the Greeks. Over the centuries, however, the visual aspects of geometry became downgraded in favour of symbolic reasoning – to the extent that Lagrange boasted in the introduction to his *Mécanique Analytique* that it contained no pictures whatsoever. One reason for this trend was the growing understanding that pictures often contain unstated assumptions. A Euclid-style diagram of a triangle cut by a line, for instance, begs the question whether it is possible to prove that the line actually meets the triangle. In fact, using only Euclid's axioms, this is not always possible. By the 1850s, however, events were set in train that would lead to a renewed emphasis on the visual, while taking on board the realization that diagrams could contain hidden traps.

BOTTLES AND BANDS

At the beginning of the 1900s, topologists obtained a complete understanding of two-dimensional surfaces – at least, surfaces without any edges that were 'closed' in the sense that they don't extend to infinity. Such surfaces fall into two categories: 'orientable' types which are spheres with a number of handles stuck on, and 'non-orientable' types like the Klein bottle, which twists around in such a manner that its inside surface merges smoothly with its outside. The best-known non-orientable surface is the Möbius band, named after Augustus Möbius (1790–1868) which has a single edge.

The Klein bottle has no edges and it is impossible to make a real physical bottle. However this is no obstacle to a topologist, who works theoretically. He can either work with the Klein bottle as an intrinsic object, not embedded in any surrounding space, or embed it (without self-crossings) in four-dimensional space. By this curious phrase mathematicians mean something entirely prosaic: the set of quadruples (x, y, z, w) of real numbers. Spaces of dimension 5, 6, and so on are defined in the same manner. They have their own 'geometry', defined through algebraic analogies with space of two and three dimensions.

**WILLIAM ROWAN
HAMILTON**
1805–65
Hamilton was the brilliant
Irish mathematician whose
algebraic study of optics
provides the basis of
quantum mechanics today.

BELOW Hamilton's brilliant
insight was to link mechanics to
geometry by constructing
equations for the path of light
rays. In his study of the patterns
produced by reflected light he
came to the fundamental
discovery of the law of least
action, which shows that light
travels along the line which
involves the least action.

Ironically, the first steps in this direction came from the development of Lagrange's analytic view of mechanics by William Rowan Hamilton (1805–65). Hamilton was a child prodigy who, by the age of ten, could understand 15 languages. In 1827 he was elected Professor of Astronomy at Trinity College, Dublin. He began by doing for optics – the mathematics of light rays – much what Joseph Louis Lagrange had managed to do for mechanics. Then he realized that both of these areas of physics shared a deep connection at a mathematical level – the same abstract formalism, but interpreted differently.

From this unification emerged a viewpoint that has since become fundamental to much of mathematical physics. It focuses on a quantity now called the Hamiltonian, and constructs dynamical or optical equations from this quantity in a systematic manner. Optics has an overt geometric aspect: the geometry of light rays; by transferring this structure to mechanics, the motion of bodies acquires geometric significance. Hamilton probably thought he was introducing algebra into optics, but the long-term consequence of his work has been to transfer geometry into mechanics. This is nowadays known as 'symplectic geometry', a term introduced by Hermann Weyl around 1910 where symplectic is the Greek word for complex.

The link that exists between geometry and core mathematics was strengthened by Felix Klein (1849–1925). Klein surveyed the disparate types of geometry that existed in his day – Euclidean, non-Euclidean, projective, inversive, whatever – and pointed out that they had a deep common structure. At the heart of geometry lay not shapes, but transformations: ways of moving shapes around. The key to Euclid was not triangles, but motions of triangles: the rigid motions that preserve the shape and size of the triangle. Geometry, said Klein, is about systems of transformations – groups in Galois's sense – and what is significant in a geometric shape is whatever remains invariant, unchanged, by the relevant group of transformations. Only Georg Bernhard Riemann's geometry of curved spaces failed to fit into Klein's ingenious scheme.

Klein proceeded to reduce large parts of the mathematics of his day – such as complex analysis, and even the solution by analytic means of the quintic equation – to geometry in this sense. His work was taken up by the Norwegian Sophus Lie (1842–1899), and group theory became the common basis of nearly all mainstream mathematics.

HILBERT'S HIT LIST

Geometry also played a central role in the work of one of the greatest mathematicians at the turn of the century, the German mathematician and physicist, David Hilbert (1862–1943). Hilbert was Klein's successor at Göttingen University, and his life's work falls into four distinct phases. In the first phase he took up the question of invariants from an algebraic viewpoint, cracked the biggest problem wide open, and abandoned the area. Next he tackled number theory, making equally dramatic strides. From there he moved on to geometry, apparently forgetting his number-theoretic ideas almost completely, and produced the first logically adequate system of axioms for Euclid's geometry. Finally he moved into mathematical physics, laying down techniques ('Hilbert space') that not so many years later were to form the basis of quantum theory.

In 1900 Hilbert attempted to summarize the mathematics of the nineteenth century, and point the way to that of the twentieth, in a famous address to the International Congress of Mathematicians. Here he stated 23 major problems whose solution, he indicated, would have a major effect in moving mathematics forward. From that time on, the most effective way to make one's reputation as a mathematician was to polish off one of Hilbert's problems. Most have since been solved, save for a few worthy exceptions including the Riemann hypothesis. The entire development of twentieth-century mathematics was strongly influenced by Hilbert's mathematical crystal-gazing.

On the other hand, Hilbert failed to anticipate many of the new theories that would spring into being after 1900. At least two of those theories were brought into existence by one person, the Frenchman Henri Poincaré (1854–1912). These were topology and the qualitative theory of dynamical systems. The former is often characterized as 'rubber sheet geometry', or the study of properties of shapes that remain

ABOVE **In attempting to solve the classic problem of whether the Solar System is stable, Poincaré discovered the theory of chaos.**

unchanged by continuous deformations: properties such as connectedness, knottedness, and the presence or absence of 'holes'. The latter led to the recognition of chaos in dynamics and the modern theory of non-linear systems, which is revolutionizing the whole of theoretical and applied science.

In topology, Poincaré developed many of the tools that are still fundamental to the subject, especially the use of abstract algebra to characterize knots, holes, and other basic topological concepts. He was something of an intuitive mathematician, and his proofs were often – to be blunt – sloppy. So good was his intuition, however, that his sloppiness usually led to creative results. Poincaré intuited his way to the answers and the key concepts; rigorous proofs could be left to other, lesser mortals. One of the main unsolved problems in modern topology, the Poincaré conjecture, which arose from an error that he made.

The conjecture states that any topological space that has certain properties in common with the three-dimensional analogue of the surface of a sphere is identical to that analogue. Calling it a 'conjecture' is over-polite: Poincaré began by tactily assuming it was obvious, then realized it wasn't and posed it as a question.

The discovery of chaos was also triggered by a mistake. Poincaré was trying to win a mathematical prize offered in 1887 by King Oscar II of Sweden. The prize question was to determine whether the Solar System is stable – the classic many-body problem. Poincaré tackled a simplified version, the 'reduced three-body problem' which was in effect the motion of the Sun, the Earth, and a speck of dust so tiny that it has no influence on the other two bodies, although they do influence it. Poincaré's prizewinning essay was published in *Acta Mathematica* in 1890, and in it he observes that under certain conditions the orbit of the dust particle can be complex and counter-intuitive – or what we would describe as 'chaotic'.

UNSOLVED PROBLEMS

There are still plenty of unsolved questions in number theory. Think of a number, say 7. If it is odd (as here) then multiply it by 3 and add 1 (to get 22). If it is even (as here) divide by 2 (to get 11). Repeat indefinitely. In this case the sequence of numbers goes 7, 22, 11, 34, 17, 52, 26, 13, 40, 20, 10, 5, 16, 8, 4, 2, 1, 4, 2, 1, 4, 2, 1... Do you always end up with the repeating cycle 4, 2, 1 wherever you start? Nobody knows. They do know that it's true for all numbers up to 700 thousand million. If you think it's obvious that the sequence always gets down to 1 and then repeats, try a variation in which for odd numbers you treble them and subtract 1. Work out what happens if you start with 17. You'll get a real surprise.

Another famous unsolved problem is the Goldbach conjecture. Christian Goldbach was an amateur mathematician who lived between 1690 and 1764. He asked his friend Leonhard Euler a deceptively simple question: is every even number bigger than two a sum of two primes? Euler couldn't solve it, and nobody else has been able to either. The answer is presumably 'yes', but we can't be sure either way. The best that is known is that every even number from some point on is either the sum of two primes, or the sum of a prime and a number that is a product of exactly two primes.

THE TRIUMPH OF ABSTRACTION

The most characteristic feature of the first two-thirds of the twentieth century was a rapid and apparently relentless drive towards increasing abstraction. The 'deep structure' of mathematics – the 'objects' under consideration – become far more complex. No longer just numbers, or sets of numbers, or mappings between sets, as in Cantor's work – but sets of mappings, mappings between sets of mappings, whatever. Increasingly, these objects were characterized axiomatically, by a list of their properties. What things *are* became unimportant: what counted is what they *do*. This approach peaked in the 1960s with the work of 'Nicolas Bourbaki' – the pseudonym of a group of young mathematicians, mostly French, who rewrote mathematics from the ground floor up in axiomatic and very abstract terms.

Axiomatic systems fall into two broad types: those that characterize specific, individual objects – such as the natural, real or complex numbers – and those that characterize a general class of objects, such as a group, ring, field, vector space, metric space, topological space, manifold (a multi-dimensional analogue of a curved space) and so forth. Those of the former kind serve to encapsulate the essence of the relevant object. The others have a tremendous virtue: generality. But they also have a vice, which is also generality.

The advantage of such an axiom system is that any theorem that follows logically from the axioms is true for all things that satisfy the axioms. It doesn't need to be reproved in each special case. This represents a substantial conceptual saving, purchased at the price of a less direct link to anything remotely concrete. By the 1960s the whole of mathematics was routinely based on various systems of axioms. But what really counts in mathematics is not your point of view: it's what you do with it. The twentieth century created a world of mathematical riches, the like of which the world had never seen before. Sadly, you have to be specialist mathematician to appreciate them.

Paramount among those riches was Poincaré's brainchild, topology. As the twentieth century

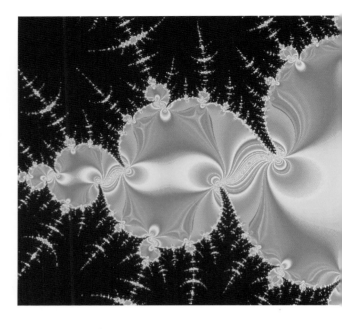

progressed, topologists made enormous progress understanding spaces of high dimensions, especially five or more. Spaces of three and four dimensions proved more elusive – as did knots, which can be viewed mathematically as closed curves embedded in three-dimensional space. But in the 1980s, New Zealander Vaughan Jones, discovered a brilliant method for distinguishing knots mathematically, the 'Jones polynomial'. Practical applications include the cutting of DNA molecules by enzymes.

The abstractions of the 1960s have been developed to the point where they provide useful payoff in many areas of science. Meanwhile, the increasingly diverse requirements of applied science and technology have led to unexpected applications of old areas and the creation of entirely new ones. An example of the former is the use of classical number theory to design concert halls with better acoustics, in which acoustic tiles are arranged in number-theoretic patterns to optimize the sound. Number theory is also used to map the surface of Venus by radar from Earth, because it invests the signal with an enormous amount of redundancy; even if only one pulse in a million is picked up on its return to Earth, it still carries information that can be built up, slowly but surely, over many months.

CHAOS, FRACTALS AND COMPLEXITY

Albert Einstein asserted that God does not play dice; that the Universe is governed by precise laws rather than chance. The area popularly known as chaos theory sheds new light on such questions, revealing that – paradoxically – precise laws may generate apparent randomness. As a result, our view of determinism, predictability, and complexity are all back in the melting-pot.

Why are tides predictable, but weather not? Tides are caused by the gravitational pull of the Sun and Moon; weather by the way the atmosphere is stirred by the sun's heat. The law of gravitation is by no means simple, so why can we predict tides years ahead, yet get the weather wrong after a few days? The answer is that the dynamics of weather quickly mixes things up, whereas the dynamics of tides gets mixed up only on scales of millions of years.

Chaotic mixing leads to unpredictability. The states of the system are continually stretched apart, and folded back in the same confined space – mixing like kneading mixes bread dough. Any system of this kind will exhibit chaos – apparent randomness in a deterministic system. The difference between a chaotic system and a non-chaotic one is simple. In non-chaotic systems, errors in initial states do not grow rapidly. In

chaotic systems, they do. In a chaotic system, at some 'prediction horizon' in the future, the error becomes larger than the true prediction, and from that moment on the original prediction bears no useful relation to the actual behaviour.

Although chaos is unpredictable, it also possesses strong elements of stability, as can be seen by using Poincaré's idea of representing dynamics geometrically. The set of states forms the 'phase space' of the system. As time passes and an initial point changes state, that point moves through phase space, describing a curve (continuous time) or sequence of points (discrete time) called its orbit. Many systems possess an 'attractor': a geometric object in phase space that apparently attract the orbits of all points that start nearby. A steady state is a point attractor; the attractor corresponding to a periodic orbit is a closed loop or cycle. Chaotic attractors are usually fractals – geometric shapes with fine structure on all scales. The concept of a fractal was developed as a systematic theory by the Polish mathematician Benoit Mandelbrot, from the 1960s onward.

ABOVE **Chaos theory explains why it is possible to predict the tidal rise and fall of the sea created by the turning of the earth beneath the moon – but not the weather for more than a few days ahead.**

BOOLE'S CONTRIBUTION

George Boole (1815–1864) was an English mathematician who became a professor at Queen's College, Cork. Boole set out to devise an algebraic formalism for logic by replacing logical statements by sets – collections of mathematical objects. To the statement 'n is a prime number' there corresponds the set that consists of all n that render the statement true – that is, the set of all primes. So logical statements can be replaced by statements about sets. Relations between statements translate into relations between sets. If P and Q are statements corresponding to sets p and q, then the logical statement 'P implies Q' is equivalent to 'p is a subset of q' – that is, every member of the set p is a member of q. Boole denotes the empty set (no members) by 0 and the universal set (everything) by 1. The intersection of two sets x and y (the set of all members that they have in common) is denoted xy, the union (what you get by combining their members into a single set) is denoted x + y, and so on. Boole worked out the algebraic laws for these operations: there are some surprises, including $x + yz = (x+y)(x+z)$, which does not hold in ordinary algebra. His idea is now known as Boolean algebra, and its most important application is to the logical basis of electronic computers.

ANDREW WILES

Andrew Wiles is the English mathematician who finally cracked the problem that had baffled mathematicians for over three centuries – proving Fermat's last theorem. His first solution, presented in 1993, proved to be false but within a year he had come up with a subtle and sophisticated solution that is now widely accepted.

RIGHT **Mathematics is the extraordinary tool that has allowed us to grapple with the fundamental secrets of the universe. The astonishing conceptual leap that allows scientists to be confident that the universe began with a Big Bang would have been impossible without mathematics.**

There are many applications of chaos. Chaos caused by Jupiter's gravitational field can fling asteroids out of orbit, towards the Earth. Disease epidemics, locust plagues, and irregular heartbeats are more down-to-earth examples of chaos. Chaos in turn is part of one of the major growth areas of current mathematical research. A striking feature of chaos is that it generates very complex behaviour from simple rules. Science tries to infer laws of nature from observations of their consequences. Chaos demonstrates that the observations may look complicated, even though the laws that lie behind them are simple. This encourages us to seek simplicity within apparently complex data. So chaos teaches an important lesson for the whole of science.

THE PROOF OF FERMAT'S LAST THEOREM

Until 1994, one of the most notorious unsolved problems was Fermat's last theorem, dating from about 1650. Recall that the lawyer and brilliant amateur mathematician Pierre de Fermat wrote about it in the margin of his copy of Diophantus' *Arithmetica*. In modern notation, Fermat's assertion was that the equation $x^n + y^n = z^n$ has no non-zero integer solutions x, y, z, whenever n is an integer ≥ 3. In other words, it works if $n = 2$, but not for any integers larger than 2.

Fermat proved his conjecture was true for $n = 4$, Euler did the same for $n = 3$, and Peter Lejeune Dirichlet and Adrien-Marie Legendre disposed of the case $n = 5$. Ernst Kummer developed the algebraic theory of 'ideals' in order to extend the range of values for which the theorem can be proved. By 1990 these methods, plus computer assistance, had allowed Joe Buhler, Richard Crandall, Tauno Metsänkylä, and Reijo Ernvall to prove Fermat's last theorem for all $n \leq 4,000,000$. That may seem strong evidence, but it doesn't prove that Fermat was right for every n. Indeed, since infinitely many integers are bigger than 4,000,000, it's not actually clear that it is strong evidence. So a complete answer looked as far off as ever.

Then, in 1993, Andrew Wiles delivered a series of three lectures at the Isaac Newton Institute in Cambridge – a new international research centre for mathematics. During the lectures Wiles revealed that he had solved a highbrow problem called the Taniyama-Weil conjecture. This grabbed the audience's attention, because they all knew that one consequence of the Taniyama-Weil conjecture was Fermat's last theorem.

Unfortunately, when Wiles sent his proof for publication, a serious gap in the logic emerged; but with the help of a colleague, Richard Taylor, and a year's hard work Wiles closed the gap and proved Fermat's conjecture to be right.

ONWARD AND OUTWARD

Mathematics is now firmly in its Golden Age. It is growing faster than ever before: its ancient riddles are being resolved at a greater rate than ever, and its relevance to the rest of science and culture is more extensive than ever before. It is being enriched by the experimental power of the computer and the conceptual tools available as software. Because of these reasons, and its intimate involvement with technology, especially information technology, mathematics has become embedded in the very fabric of human lives.

Without it, we are lost.

With it, we shall reach for the stars.

VOLUME OF SPHERE

AT A GLANCE

1500 BC
Babylonians develop a number system.

600 BC
Thales develops proofs for his theorems.

530 BC
Pythagoras devises his theorem about right-angled triangles.

TETRAHEDRON

300 BC
The Hindus develop their Brahmi number symbols.

300 BC
Euclid writes his *Elements of Geometry*, perhaps the most influential book in the history of mathematics and the basis of all geometry until the nineteenth century.

220 BC
Archimedes develops ways of measuring the volumes of solids such as spheres.

EUCLID

200 BC
Apollonius analyzes sections of the cone – the ellipse, parabola and hyperbola.

AD 470
Tsu Ch'ung Chi establishes π to nine decimal places.

AD 662
By this date the Hindu number system has developed into today's decimal system.

AD 825
Al-Khawarizmi writes the first major treatise on algebra.

5x-10=0

1220
Leonardo of Pisa introduces Hindu-Arabic numbers to Europe, and invents Fibonacci numbers.

PYTHAGOREAN TRIPLES

1520
Scipio del Ferro solves all three types of cubic equations.

1572
Raffael Bombelli suggests that negative numbers might be useful.

1637
René Descartes creates coordinate geometry.

1637
Pierre de Fermat poses one of mathematics' most enduring problems – Fermat's last theorem – and suggests that he has a remarkable solution which he fails to reveal.

ISAAC NEWTON

1665
Isaac Newton develops binomial theorems and discovers how to use calculus.

1770
Independently of Newton, Wilhelm Leibniz also discovers calculus.

1730
Leonhard Euler develops a huge number of theorems and puts trigonometry and differential calculus on a firm basis.

1755
Joseph-Louis Lagrange begins his great book *Analytical Mechanics* in which he uses the calculus of four-dimensional space to solve mechanical problems.

PIERRE DE FERMAT

1801
Karl Gauss writes his book *Researches in Arithmetic* which provides the basis of modern number theory.

1837
William Hamilton links geometry and mechanics with his work on the mathematics of optics.

WILLIAM HAMILTON

1832
Évariste Galois employs Lagrange's ideas about permutations to establish 'Galois groups' and reveal that algebra can be used for establishing general truths.

1854
Georg Riemann introduces the idea of Riemann surfaces and multi-dimensional space.

1880
Georg Cantor develops a highly original system of arithmetic for infinity, and subsequently his ideas on the theory of sets of points become crucial to analysis and topology.

1884
Christian Klein finds the way to unify both Euclidian and non-Euclidian geometry.

1900
David Hilbert devises the idea of 'Hilbert space' – the mathematics that underpins quantum theory – and outlines 23 major mathematical challenges.

1906
Henri Poincaré develops the mathematics of topology and qualitative theory of dynamic systems.

TOPOLOGY

1962
Lotfi Zadeh invents fuzzy logic.

1964
Benoit Mandelbrot discovers fractals.

1994
Andrew Wiles provides proof of Fermat's last theorem.

MÖBIUS BAND

Energy and Motion

PHYSICS

Physics is the king of sciences, the rock upon which all of science is based. It is said – with some truth – that chemistry is applied physics, and that biology is applied chemistry. Physics itself, of course, is underpinned by mathematics; but this is a different kind of situation. Mathematics is based on laws which can be proved by logic and reason – if you like, by philosophy. You do not carry out experiments in mathematics to determine, say, that the angles of a triangle add up to 180 degrees. You prove it by logic and reason, and it is an absolute truth. If you measure the angles of a triangle drawn on a flat piece of paper and the total comes out as something other than 180 degrees, it is the 'experiment' that is wrong, not the mathematical truth. In physics, though, the experiment is the ultimate test; as Richard Feynman once said, 'if it disagrees with experiment, then it is wrong'. No physics theory, no matter how elegant and appealing, can be regarded as a good description of the way things behave in the real world if it makes predictions that disagree with the outcome of experiments.

GALILEO GALILEI
1564–1642
The first scientist of the modern era. Galileo introduced the idea of rigorous experiments to test scientific ideas.

BELOW **The scientific image of the Universe that emerged from the work of people like Galileo and Newton resembled a machine, running like clockwork according to a well-determined pattern.**

The first person who fully appreciated this was Galileo Galilei, who lived from 1564 to 1642. There is no doubt that physics as we know it (indeed, science as we know it) was born in the seventeenth century. You can argue that the publication of Galileo's greatest work *Two New Sciences* in 1638 was the key date, or maybe the publication of Isaac Newton's greatest work *Mathematical Principles of Natural Philosophy* (usually known, from its Latin title, as the *Principia*) in 1687 was the moment when physics came of age. But there can be no argument that before Galileo there was no such thing as true science, and that after Newton physics was fully established in its recognizable modern form. So all the achievements of science have been made in a little over 300 years.

ABSTRACT PHILOSOPHIZING

What passed for science before Galileo was a kind of abstract philosophizing, the ridiculousness of which is highlighted by the story of the weights being dropped from the Leaning Tower of Pisa. In fact Galileo himself never carried out such an experiment. He investigated the way things fall under the influence of gravity by rolling different balls down inclined planes, and timing their descent using the most accurate techniques possible at the time.

But the point is that there had been a debate among the 'natural philosophers' about whether a light ball and a heavy ball dropped from a height would fall at the same speed and reach the ground together, or whether the heavier object would fall faster and reach the ground first. Before Galileo, none of these 'philosophers' had bothered to carry out the experiment; instead, they tried to find an answer by pure reason, the way you would try to prove a mathematical theorem. It was Galileo who began to test theories by actual experiments, and it was in the wake of Galileo's assertion – based on his experiments – that the heavy and light objects would indeed fall at the same rate that one of his opponents actually carried out the experiment from the Leaning Tower. He was attempting to prove Galileo wrong, but he succeeded only in proving him right: the two weights hit the ground at the same time. This is an important feature of an honest experiment – it always tells you the way the world really is, no matter how fervently you are hoping for a different answer.

UNIVERSAL LAWS

Isaac Newton, who lived from 1642 (the year Galileo died) to 1727, was also a superb experimenter who fully understood the need to test theories in the way pioneered by Galileo. But he also realized that the behaviour of objects in the world at large is governed by laws that are literally universal: they apply to everything on Earth and in the Universe, at all times.

Most famously, Newton realized that the force of gravity which pulls an apple from a tree is the same force, obeying the same law, as the force of gravity which holds the Moon in orbit around the Earth, and the planets in orbit around the Sun. Indeed, the law of gravitation is another of the universal laws that applies to every object in the Universe, including the stars themselves.

It was Newton who, by his rigorous and logical approach to science, destroyed the idea that the Universe might be governed by capricious gods who could determine the fall of an apple or the motion of a star at their whim. He replaced this by the concept of a Universe running inexorably in accordance with predetermined, inviolable laws: laws of nature which could be discovered, if physicists were skilful enough to carry out the appropriate experiments. And that has been the story of physics for the past three centuries – the search for universal laws, which can be revealed by carrying out a series of ever more subtle experiments.

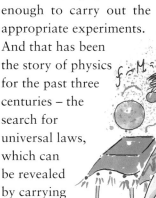

ABOVE **The most famous demonstration of Newton's laws was carried out by astronauts on the Moon who dropped a hammer and a feather together. Since there was no air to impede the descent of the feather, they both hit the surface at the same time.**

CENTRE **Newton said that any object free from outside influences will move in a straight line through space. Planets move in closed orbits because they are tugged out of their default straight-line course by the gravitational pull of the Sun.**

LEFT **Isaac Newton is sometimes called 'the last sorcerer', because he also carried out experiments in alchemy. To most of his contemporaries, his ideas about physics involved just as much magic as the idea of turning lead into gold.**

LEFT **If you've ever tried to get a heavy grass roller moving, you'll know exactly what inertia is. No object will move** unless it is pushed or pulled by a force – and the heavier an object is, the greater the force needed to get it moving.

INERTIA

One of the most important scientific insights is that any object keeps moving in a straight line at constant speed (or stays still) unless it is pushed or pulled by a force. This was first formulated as a law of nature (his first law of motion) by Isaac Newton. This property, inertia, was not obvious, and even Galileo had only got halfway to understanding it. Galileo realized, from his experiments rolling balls down inclined planes, that if there were no friction a ball would keep rolling towards the horizon. But he also knew that the Earth was round, and that motion 'towards the horizon' was actually motion along part of a circle around the Earth. So he thought that the natural law of inertia was for things to move in circles, which would explain the orbits of the planets. It was Newton who realized that, like everything else, the planets 'want' to move in straight lines, but are pulled off course by a force – the force of gravity.

The neatest demonstration of inertia was given in 1640, two years before the birth of Newton. The French polymath Pierre Gassendi decided to see whether an object dropped from a moving vehicle would fall straight down, or would carry with it the forward motion of the vehicle. He had a slave galley rowed flat out across the calm Mediterranean, and climbed to the top of the mast and dropped a series of balls. They fell at the foot of the mast, having shared in the ship's forward motion even while they were falling.

ABOVE **Newton realized that the force that pulls an apple off a tree is the same as the force which holds the Moon in orbit around the Earth – gravity. Only an inverse square law of gravity explains the motion of both the apple and the Moon.**

ORBITAL THEORY

Newton said that he obtained his insight into the nature of gravity by considering the fall of an apple and the orbit of the Moon. He knew that the acceleration caused by gravity will make any object near the surface of the Earth, including an apple dropping from a tree, fall by 16 feet (in the units that he used) in the first second that it is falling. According to his first law of motion the Moon 'wants' to move in a straight line, so there must be a force tugging it sideways to keep it in orbit around the Earth. Imagine that the Moon moves forward in a straight line for a second, then receives a sideways tug putting it back on its circular orbit once again. If this is continuously repeated, and if all the tugs are the right size, the Moon will indeed stay in orbit. Newton's insight was to realize that the source of this tugging is the same as the source of the force that pulls an apple from a tree to the ground – the Earth's gravity.

The repeated nudges needed to keep the Moon in its orbit amount to a sideways shift of $1/20$th of an inch, in the units used by Newton, every second. So the Moon 'falls' under the influence of the Earth's gravity by $1/20$th of an inch in one second. This is $1/3800$th of the amount the apple falls near the surface of the Earth. But the Moon is 60 times further away from the centre of the Earth than the apple is. And 3800 is just over 60 squared, so the force of gravity felt by the Moon is weakened by the square of its distance. Gravity obeys an inverse square law, and the same law explains the orbits of the planets around the Sun.

A rock with the same mass as the Moon

ISAAC NEWTON
1642–1727
The man who established the scientific method, Newton is generally regarded as the greatest scientist of all time. He introduced the idea of universal laws which apply everywhere in time and space.

A rock with the mass of an apple

NEWTON'S OTHER LAWS

Newton's second law of motion says that when a force is applied to an object moving in a straight line at constant speed, its velocity changes at a rate proportional to the force that is applied, and in the direction that the force is applied. In other words, it accelerates. But the bigger the mass of the object, the harder it is to make it accelerate, because it has more inertia. In mathematical language, if the force that is applied is *F* and the mass of the object is *m*, the acceleration *a* is given by the equation $F \div m = a$.

The third law is the one exploited by rockets when they climb up from the Earth's surface. Whenever a force (what Newton called an action) is applied to an object, the object pushes back with an equal and opposite force, the reaction. As you stand on the ground, gravity is trying to pull you towards the centre of the Earth. But you do not accelerate downwards, because the ground is pushing up with an equal and opposite reaction, cancelling out the effect of gravity so that you stay still.

When hot gas is pushed out of the back of a rocket, the reaction pushes back on the rocket. It is this reaction that lifts the rocket upwards, against the pull of gravity. The hot gas does not have to push against the ground, the air, or for that matter anything at all, and this is why it can propel the rocket through space.

FORCE AND ACCELERATION

ABOVE **Newton's second law of motion shows why the harder the racket hits the ball, the faster it goes.**

The force of gravity pulls any object along a line passing through the centre of the object

ABOVE **The heavier an object is, the greater its gravitational attraction. So why does Gravity (the man in our picture) pull the heavy rock no faster than the light rock? The answer is that the heavy rock has greater inertia, and so takes more pulling.**

Gravity is a force that affects all objects equally, regardless of their mass

THE NATURE OF LIGHT

The laws that Newton himself discovered dominated physics for more than 200 years, and are still the basis of a great deal of everyday technology ranging from the behaviour of the balls on a pool table to the design of a spacecraft sent to probe the outer planets of the Solar System. They tell us that objects keep moving in straight lines unless they are pushed or pulled by forces, and they tell us how objects react to being pushed and pulled by forces. It was natural that Newton should try to explain light in the same way. He thought of light as being made up of a stream of tiny particles, or 'corpuscles', which bounced around (reflecting from mirrors and so on) in accordance with the same laws that describe the way an apple falls from a tree, or how two billiard balls react to a collision.

This idea dominated scientific thinking about the nature of light for well over 100 years, right through the eighteenth century, but not because there was overwhelming experimental evidence in its favour. It was because the idea came from Newton, who was regarded as the oracle on all matters scientific. This was unfortunate, because he wasn't right. Light does not travel through space like a stream of little cannon balls (as we shall see later, though, in a sense Newton was only half wrong). And, crucially, the developing understanding of light provided the key to the next great development in the scientific understanding of the world. Partly thanks to the shadow Newton cast over the eighteenth century, the improved understanding of light – and the ensuing scientific revolution – had to wait until well into the nineteenth century.

THE WAVE THEORY

It was particularly unfortunate for Christiaan Huygens, a Dutch physicist and astronomer who lived from 1629 to 1695, and whose active scientific career overlapped with that of Newton. If there had been no Isaac Newton, Huygens would be remembered as the most influential scientist of his time. He explained the motion of a pendulum, and used his understanding to design the first successful pendulum clock. A skilled astronomer, he was the first person to give a proper explanation of the rings of Saturn. But his greatest achievement was his theory of light – a theory which described light in terms of waves, not particles. Huygens' wave theory was first put forward in 1678, and published in complete form in 1690. But it was largely ignored for well over 100 years, simply because it was incompatible with Newton's theory (or so it seemed at the time). Huygens treated light as vibrations in some all-pervading medium that filled the Universe, and his model explained both refraction and reflection. Newton's corpuscular model also explained refraction and reflection, but the two explanations were based on different premises.

According to Huygens' theory, light must travel more slowly in a denser medium (and therefore more slowly in water than in air), while according to Newton light must travel more quickly in a denser medium (and therefore faster in water than in air). The test could not be carried out accurately enough to distinguish between the rival ideas until well into the nineteenth century. When it was finally tested, Huygens was vindicated. But by then it had already been established beyond reasonable doubt that light (and all electromagnetic radiation) does indeed travel as a wave.

CHRISTIAAN HUYGENS
1629–93
Unlucky to have his career overshadowed by the towering achievements of Isaac Newton, Huygens was the first person to develop a satisfactory wave theory of light.

LEFT **When white light, or light from the Sun, is passed through a glass prism, it is split into all the colours of the rainbow. We now know that each colour corresponds to a different wavelength of light.**

REFLECTION AND REFRACTION

The first wave theory of light was developed by **Christiaan Huygens** in the late seventeenth century. He came up with a model to describe how the waves spread, which is still useful today.

According to Huygens, each point on the leading edge of an advancing wavefront can be considered as a new source of waves, and the surface joining these 'secondary wavelets' becomes the new advancing wavefront. This idea makes it possible to calculate how waves move using geometry alone, and avoids the need for a full understanding of the physical reality of what the waves are.

Reflection is very easy to understand in these terms, because light waves bounce off a reflecting surface (such as a mirror) at the same angle at which they hit it. As far as distinguishing between waves and Newton's particle theory goes, however, this is a useless test, since particles bounce off a reflecting surface in the same way, just as billiard balls normally bounce off the cushions of the table.

ABOVE **You can see refraction by simply poking a stick into clear water. It is really the light rays from the stick that are bent, not the stick itself.**

BENDING RAYS

Refraction is more interesting. When light moves from a less dense medium (such as air) into a more dense medium (such as water), the beam of light bends so that it makes a less acute (more nearly vertical) angle to the surface between the two mediums. This bending of the rays of light is the reason why a straight stick placed half in and half out of a pool of water looks as if it is bent.

The wave theory explains this phenomenon in terms of the light moving more slowly in the denser medium, but according to Newton's corpuscular theory it happens because the particles travel faster in the more dense medium. The question remained unresolved until 1850, when experimental techniques were good enough for the French physicist Jean Foucault to actually measure the difference. Foucault's measurements showed that the speed of light is less in water than in air, and confirmed that light travels as a wave.

REFLECTION AND REFRACTION

RIGHT **Both the wave theory and the particle theory of light made the same predictions about how light rays are reflected. But the two theories made opposite predictions about how light rays should be affected by refraction. Experiments prove that light travels as a wave.**

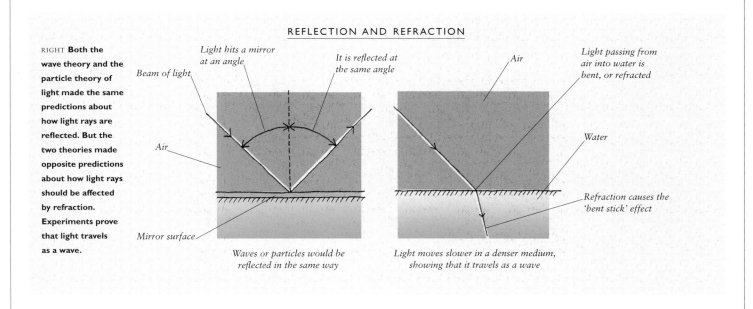

Beam of light

Light hits a mirror at an angle

It is reflected at the same angle

Air

Mirror surface

Waves or particles would be reflected in the same way

Air

Light passing from air into water is bent, or refracted

Water

Refraction causes the 'bent stick' effect

Light moves slower in a denser medium, showing that it travels as a wave

THOMAS YOUNG
1773–1829

A polymath who dabbled in science and medicine, Young was the chief figure in the decoding of the Rosetta stone. In the first years of the nineteenth century he revived the wave theory of light, which had been dormant since Isaac Newton had championed the rival corpuscular theory at the end of the seventeenth century.

WAVES OF LIGHT

The key experiments proving the wave nature of light were carried out early in the nineteenth century by Thomas Young in England (and, following his re-invention of the wave theory, by Augustin Fresnel in France). The most memorable demonstration was devised by Young, who shone light through two narrow slits cut in a sheet of card. The light spreading out from each of the two slits interacted to produce a pattern of light and dark stripes on a second card, in exactly the way that ripples spreading out on a pond interact to produce an interference pattern. There was no doubt that light was moving like a wave. But what, exactly, was 'waving'? We can see ripples on a pond, as the water moves up and down; we can hear sound waves, as air is squeezed by the passage of the waves. But what is it that ripples when a light wave passes by? Physicists gave it a name, the aether, but had no idea what it might be made of.

It took two people, in successive scientific generations, and decades of work to resolve that

ABOVE **When ripples spread across the surface of a liquid they interfere with each other and make patterns. Young's experiments with light gave similar results.**

puzzle. Part of the reason why it took so much effort was that it involved introducing a completely new concept into science, the idea of a field of force. The idea came from Michael Faraday, who was one of the greatest popularizers of science (as well as one of the greatest scientists) among other things. He started the tradition of the Christmas Lectures at the Royal Institution in London.

FIELDS OF FORCE

Faraday came up with the now familiar idea of a 'line of force', a term he invented and first used in a scientific paper in 1831. The concept is easiest to comprehend in terms of magnetism, and an experiment that is familiar to most schoolchildren. When a magnet is placed under a sheet of paper that is covered with iron filings, and the sheet of paper is tapped gently, the filings arrange themselves in curving lines between the north and south poles of the magnet. These lines mark lines of force: the trajectories that would be followed by a free single magnetic pole (if such a thing existed) travelling in space from one pole of the magnet to the other.

Lines of force can be generated by a permanent magnet, or by an electric current passing through a coil. The force field itself can also generate an electric current. In a document that he wrote in 1832, but showed to nobody at the time, Faraday suggested that light might be explained in terms of the vibration of these lines of force, and that it would take a definite time for any electric or magnetic influence to spread from a magnet into the world at large if the magnet was disturbed in some way. The idea of fields of force came to be one of the most crucial in science.

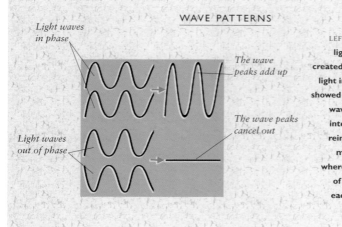

WAVE PATTERNS

Light waves in phase

The wave peaks add up

Light waves out of phase

The wave peaks cancel out

LEFT **How the pattern of light and dark stripes is created by the two beams of light in Young's experiment showed that light behaved as waves. Where the waves interacted in phase they reinforced each other to make light stripes, and where they interacted out of phase they cancelled each other out to make dark stripes.**

LINES OF FORCE

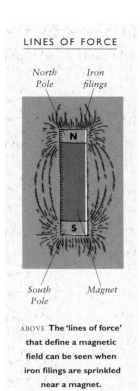

North Pole   *Iron filings*

N

S

South Pole   *Magnet*

ABOVE **The 'lines of force' that define a magnetic field can be seen when iron filings are sprinkled near a magnet.**

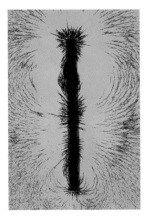

ABOVE **The filings are attracted to the north and south poles arranging themselves in curving lines between the poles – the lines of force. Faraday suggested that light could be regarded as waves running along these lines of force.**

THE EXPERIMENT WITH ONE HOLE

You can actually demonstrate the wave nature of light for yourself, using your fingers. Hold your hand up to a light, with the palm towards you and the fingers touching. Now move a pair of fingers very slightly apart (only very slightly). In the narrow gap between the two fingers, you will see a couple of dark lines. These are caused by interference as the light waves bend round the edges of the narrow slit you have made with your fingers and interact with each other.

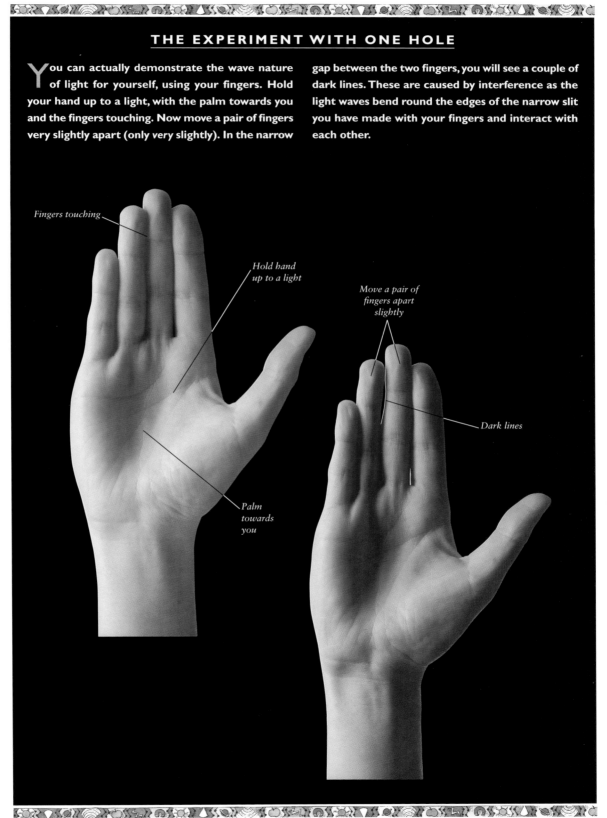

Fingers touching

Hold hand up to a light

Move a pair of fingers apart slightly

Dark lines

Palm towards you

59

THE EXPERIMENT WITH TWO HOLES

All of the mystery and strangeness of the quantum world is contained in the double slit experiment, originally used by Thomas Young to prove that light is a wave. More recently, Richard Feynman has said that this experiment 'has in it the heart of quantum mechanics', and on another occasion that 'any other situation in quantum mechanics, it turns out, can always be explained by saying 'You remember the case of the experiment with two holes? It's the same thing.' So here it is.

The experiment is designed to demonstrate the behaviour of waves, and everybody knows what a wave is – we have all seen ripples in a pond, or in a bathtub. If you drop a pebble into a still pond, you can see waves spreading out neatly in circles around the place where the pebble fell. If you drop two pebbles into a pond, you get two sets of waves, which interfere with each other to make a more complicated pattern. What Young did was find a way of making this more complicated pattern with light.

SPREADING WAVES

To get a uniform source of light, he first passed it through a single hole in a screen, so that waves spread out from this single hole like ripples from a single stone dropped in a pond. It was these spreading waves that encountered the screen with two holes cut in it (Young used very narrow slits cut with a razor, but tiny round holes, like pinpricks, work just as well).

Because the light came originally from the single hole in the first screen, the waves spreading out on the far side of the second screen from each of these two holes marched in step with one another, like ripples made by two stones dropped in a pond at precisely the same instant. The circular waves (or strictly speaking, semicircular) spread out on the other side of the screen with two holes, and fell upon a third screen. This screen was kept completely dark, apart from the light created by the waves radiating from the two holes.

RICHARD FEYNMAN
1918–1988
Arguably the greatest physicist since Newton, Feynman developed the most complete and straightforward version of quantum theory in the 1940s – as his doctoral thesis. He went on to receive the Nobel Prize for his theory of quantum electro-dynamics, and made many other major contributions to physics.

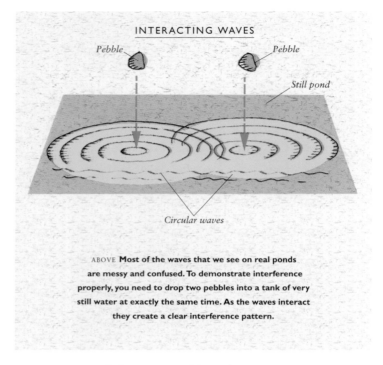

INTERACTING WAVES

Pebble

Pebble

Still pond

Circular waves

ABOVE **Most of the waves that we see on real ponds are messy and confused. To demonstrate interference properly, you need to drop two pebbles into a tank of very still water at exactly the same time. As the waves interact they create a clear interference pattern.**

PATTERNS OF LIGHT AND DARK

When Young shone light through his apparatus he produced a pattern of light and dark bands on the third screen. In places where the two sets of waves happened to be both at a peak, they added together to make a bright band of light; but in places where the peak of one wave coincided with the trough of the other wave, they cancelled out to create a dark band.

The pattern of light and dark bands on the final screen was exactly the pattern that ought to be produced by the interference of waves. Indeed, you could use the measured spacing of the light and dark bands on the screen to calculate the wavelength of the light passing through the experiment. This is why, after Young's work, nineteenth-century physicists were convinced that light is a straightforward wave phenomenon – but since the wavelength of light revealed in this way is only about half a millionth of a metre, it also explains why nobody had noticed it before.

It's crucially important to notice that the interference pattern is not the same pattern that you would get if you add up the light from each of the two holes independently. If just one of the holes in the middle screen was open, you would get a patch of light on

the final screen centred behind that hole. If the other hole were open instead, you would get a patch of light behind that hole. But when both holes are open, creating the striped effect, the brightest patch on the final screen is exactly halfway between the two bright spots you would get if each hole were opened separately.

Unfortunately, though, there is strong evidence that light can also behave like a stream of tiny particles, called photons. So light seems to be both a particle and a wave. This 'quantum duality' is brought home forcefully when the equivalent of Young's slit experiment is carried out using electrons, which are usually thought of as particles. This really has been done, quite recently, by a team from the Hitachi research labs and Gakushuin University in Tokyo.

PARTICLES BEHAVING AS WAVES

In this version of the experiment with two holes, the electrons were fired through the apparatus one at a time, from the tip of an electron microscope. There is no doubt that they left the 'gun' in the form of individual particles. On the far side of the detector screen, their arrival was recorded on a screen like a TV screen, where each electron made a flash of light. There is no doubt that each electron arrived on the far side of the experiment as a single particle. Unlike a conventional TV, however, this detector screen kept the bright spot corresponding to the arrival of each

individual electron lit up as more and more electrons came through. What pattern would you expect to build up on the screen?

If electrons are particles, the answer seems obvious from everyday experience. If you stood on one side of a tall wall with two holes in it, and threw pebbles in the general direction of the wall, without bothering to aim very carefully, some of them would go through each hole. The pebbles would pile up in two heaps on the other side, one behind each hole. So did the Japanese team see two bright spots building up on the detector screen, one behind each hole? No!

As dozens, then hundreds, then thousands of electrons passed through this high-tech version of the experiment with two holes, one after the other, each one made its own distinct spot of light on the detector screen. Yet the pattern that built up was the striped interference pattern appropriate for waves. It was as if each electron had dissolved into a wave, gone through both holes at once, interfered with itself, and chosen its destination on the basis of its place in the interference pattern.

Even worse, in order to produce the overall pattern, the behaviour of the electrons had to be coordinated in some way. Although it seems ridiculous, each electron seemed to know its proper place in the pattern, as if it was aware of the behaviour of all the electrons that had gone before it, and those that were still to come – although this seems impossible.

Don't worry if you do not understand this. You are in good company. The American physicist Richard Feynman won the Nobel Prize for the theory of quantum electrodynamics, and he probably understood quantum theory better than most. But even he said, 'I think I can safely say that nobody understands quantum mechanics… Do not keep saying to yourself, if you can possibly avoid it, 'But how can it be like that?' because you will go 'down the drain' into a blind alley from which nobody has yet escaped. Nobody knows how it can be like that.'

INTERFERING LIGHT WAVES

As the waves cross they interact in different ways

The holes are just pinpricks

The wavelength of the light from each hole is the same

Destructive interference makes a light band

Constructive interference makes a dark band

Light and dark bands form on the screen

LEFT **Light waves start off in phase as they pass through the two holes, but the peaks and troughs of the waves interact and either cancel to create a dark band, or build on each other to create a light band.**

THE KINETIC THEORY

One of the greatest triumphs of the classical (Newtonian) model of the world was its application to the theory of gases in the nineteenth century. The theory is based on the image of atoms as tiny ball-like objects, bouncing around in a gas, colliding with one another and with the walls of their container, and moving precisely in accordance with Newton's three laws of motion. The pressure exerted by a gas on the walls of its container is interpreted as a result of billions of collisions between these tiny particles and the walls, each collision involving action and reaction in the way described by Newton.

The kinetic theory explains the properties of gases that had been discovered by experiment – in particular, the way the pressure of a gas and its temperature go up if it is squeezed into a smaller volume, and go down if it is allowed to expand into a larger volume.

HIGH-SPEED COLLISIONS

At a temperature of 0°C and a pressure corresponding to the pressure at sea level, each litre of air contains 30,000 billion billion molecules. They move at an average speed of 450 metres per second, and each molecule undergoes some five million collisions every second, either with other molecules or with the walls of its container. So the average distance that a molecule travels between collisions is just 90 millionths of a metre.

These ideas are extended into the kinetic theory of heat, which explains heat in terms of the speed with which the component atoms (or molecules) of a substance are moving. In a solid, the atoms have relatively little energy, and so do not jiggle about very much. They are held in place by the 'sticky' electromagnetic forces between

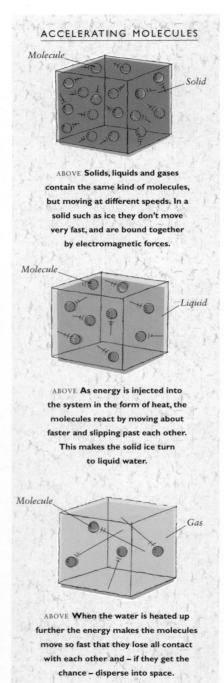

ACCELERATING MOLECULES

Molecule

Solid

ABOVE **Solids, liquids and gases contain the same kind of molecules, but moving at different speeds. In a solid such as ice they don't move very fast, and are bound together by electromagnetic forces.**

Molecule

Liquid

ABOVE **As energy is injected into the system in the form of heat, the molecules react by moving about faster and slipping past each other. This makes the solid ice turn to liquid water.**

Molecule

Gas

ABOVE **When the water is heated up further the energy makes the molecules move so fast that they lose all contact with each other and – if they get the chance – disperse into space.**

ABOVE **When liquid nitrogen is poured out of a refrigerated container, its molecules gain energy and speed up to become a gas, mingling with the air and disappearing before your eyes.**

neighbouring atoms. When the same substance exists as a liquid, at a higher temperature, the atoms jiggle about so much that they break free from a single location and can move about, slipping past one another; but they are still moving slowly enough to feel the electromagnetic forces and be partially restrained by them. In a gas, the atoms have so much energy and move so fast that there is no time for them to be restrained by the other atoms.

Different substances melt and vaporize at different temperatures because there are variations in the stickiness of atoms, and because it takes more energy to make heavier atoms move faster.

The kinetic theory leads to the idea of an absolute zero of temperature, when atoms have the minimum amount of energy and are moving as slowly as possible, coming effectively to a standstill. This point is –273°C, or zero on the Kelvin scale.

LINES OF FORCE

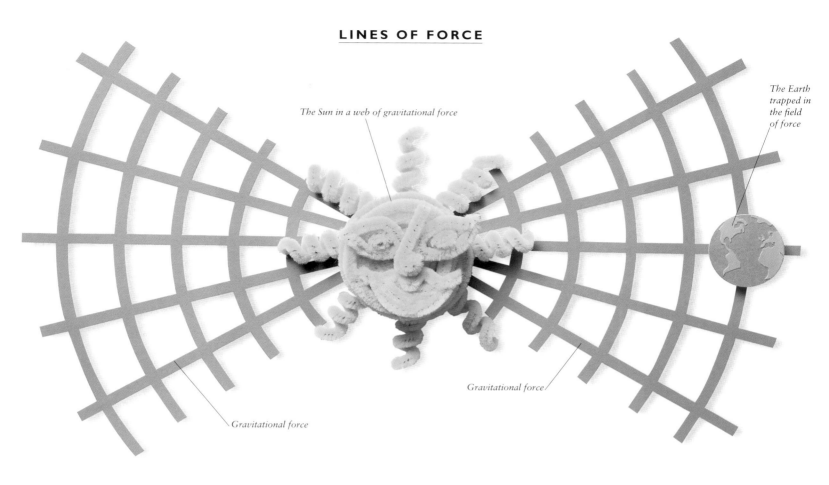

The Sun in a web of gravitational force

The Earth trapped in the field of force

Gravitational force

Gravitational force

ABOVE **Faraday thought of the Sun as sitting in a web of forces, rather like a spider. If the Earth was suddenly dropped into the web it would know the Sun was there, because it would 'feel' the web.**

Partly because of ill health, it wasn't until 1844 that Faraday was ready to go public with a full account of his force-field theory. Then, in a lecture at the Royal Institution, he provided his audience with a classic example of a 'thought experiment', making it clear that his ideas applied to all the forces of nature and not just magnetism.

Faraday asked his audience to imagine the Sun sitting on its own in space, and to picture the Earth suddenly placed alongside it, at its proper distance. He then asked the audience to consider how the Earth would 'know' that the Sun was there?

According to Faraday, a web of lines of gravitational force – the gravitational field – must extend outward from the Sun into the Universe, even if there were no planets to feel the Sun's gravitational influence. If the Earth was dropped into the field of force, it would respond instantly to the field at the place it was located. It would not have to wait for some message to travel to the Sun and back to find out that the Sun was there, and adjust its orbit accordingly.

In 1846, Faraday summed up his ideas about light in another lecture. 'The view which I am so bold as to put forth,' he said, 'considers, therefore, radiation as a high species of vibration in the lines of force which are known to connect particles, and also masses of matter, together. It endeavours to dismiss the aether, but not the vibrations.' In other words, the only thing that had to ripple when light travelled from place to place was the electromagnetic field itself.

MICHAEL FARADAY
1791–1867
Discovered the principles of both the electric motor and the electric generator, then went on to lay the foundations of the modern understanding of electromagnetism in terms of fields of force.

ABOVE **Michael Faraday's study of electricity and magnetism led to the invention of many things that seem indispensible in the modern world, like the personal stereo.**

BELOW **Our eyes can see only visible light – just a small proportion of the wide range of electromagnetic radiation, that goes from gamma rays to radio waves.**

A COMPLETE THEORY OF LIGHT

Faraday died in 1867, just short of his 76th birthday. But he lived long enough to see his ideas taken up and turned into a complete theory of light by James Clerk Maxwell, a Scot who graduated from the University of Cambridge in 1854 and spent the next ten years developing Faraday's ideas into mathematical language.

Maxwell started out by investigating the way forces are transmitted by electric and magnetic fields, and determined how fast disturbances in such fields would travel through space. The answer turned out to be the speed of light – a discovery so dramatic and exciting that Maxwell himself italicized the relevant passage in a paper published in 1862, pointing out that 'we can scarcely avoid the inference that *light consists in the transverse undulations of the same medium which is the cause of electric and magnetic phenomena*'. His excitement was caused by the way the discovery linked apparently unconnected phenomena – the way a magnet pulls on an iron nail, the way lightning is generated in a storm, and the way light reflects from a pond – in one package: the first great unification in physics.

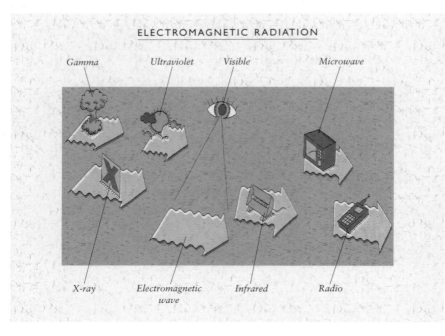

ELECTROMAGNETIC RADIATION

Gamma   Ultraviolet   Visible   Microwave

X-ray   Electromagnetic wave   Infrared   Radio

In 1864 Maxwell published 'A Dynamical Theory of the Electromagnetic Field', a paper which contains virtually everything you need to know about electricity, magnetism and light summed up in just four equations. Maxwell's equations tell you the strength of the force between any two electric charges, the power of a magnetic field generated by an electric current of a certain strength flowing in a wire, and much more. And they show that light, and all electromagnetic radiation, is a form of transverse ripple in the fields of force that fill the Universe.

ELECTROMAGNETIC RADIATION

Think of this as like sending ripples along a stretched rope by flicking it from side to side. According to the old wave theory, the rope might represent the aether. But Faraday had discovered that a changing electric field creates a magnetic field at right-angles to the electric field, and conversely a changing magnetic field creates an electric field. If you imagine electric ripples running from side to side along the rope, they will create magnetic ripples that run up and down along the rope. The magnetic ripples generate electric ripples, and so on. There is no need for a rope – the aether – because the waves are self-generating in the field of force. Once energy is put in, perhaps by jiggling a magnet to and fro, the waves continue indefinitely, locked together in accordance with Maxwell's equations.

Even better, like all good theories Maxwell's made a prediction. There should be all kinds of electromagnetic waves, not just light waves. In particular, there should be waves of longer wavelength, travelling at the same speed as light, which might be generated in the lab using electric currents in wires. Just such waves were generated and studied by Heinrich Hertz, in the 1880s; they are now called radio waves.

JAMES CLERK MAXWELL
1831–79
Maxwell took up the study of electromagnetism and light where Faraday left off, and developed a complete description of all classical electromagnetic phenomena, summed up in a set of four equations – Maxwell's equations.

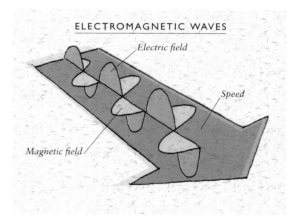

ELECTROMAGNETIC WAVES

Electric field

Speed

Magnetic field

ABOVE **An electromagnetic wave is really two waves at right angles to each other: one is the electric field and one is the magnetic field. They have the same wavelength which, for waves in the visible spectrum, is the wavelength of the light.**

CENTRE **There are few more spectacular demonstrations of the link between light and electricity than a bolt of lightning. The brief but huge electrical surge is almost blindingly bright.**

RELATIVITY AND THE SPEED OF LIGHT

By the end of the nineteenth century it was clear that electromagnetic radiation, including light, was a form of wave vibration that travelled at a certain speed, a speed that, defined by the laws of nature, could be worked out from Maxwell's equations. Between them, Maxwell's equations and Newton's laws explained everything that was known about the physical world. Or did they? There were two puzzles that weren't quite explained by the end of the 1890s, and between them they led to the biggest upheaval in physics there has ever been, producing quantum theory and the theory of relativity.

We'll take relativity first, because it follows directly from Maxwell's work on the speed of light, and because the theory was completed first – although people were already puzzling about quantum ideas before Albert Einstein produced his two theories of relativity.

The puzzle that Einstein (born in 1879, the year Maxwell died) latched on to was that Newton's laws and Maxwell's equations disagree on one crucial point. Newton's laws describe, among other things, the way things move, and the relationships between velocities. If I am driving in a car down a straight road at 60 km per hour, and you are walking towards me at 6 km per hour, then the relative speed between us is 66 km per hour (60 + 6). If I were foolish enough to fire a gun forward from the moving car, and the bullet travelled at a speed of 400 km per hour relative to me, it would zip past your ear at 466 km per hour (400 + 60 + 6).

But what if I switched on the headlights of the car, so that a beam of light travelled forward from the vehicle? How fast would the light be moving relative to me, and how fast relative to you? The speed of light is specified precisely by Maxwell's equations, and has been measured very accurately by experiments. It is 300,000 km per second (more than a billion km per hour), but is usually denoted by c. Common sense (and Newton's laws) would say that the speed of the beam of light relative to

NORMAL SPEED

The bullet's speed can vary

ABOVE **If you fire a bullet from a speeding car, it travels past a bystander at a rate that is equal to the normal speed of the bullet plus the speed of the car.**

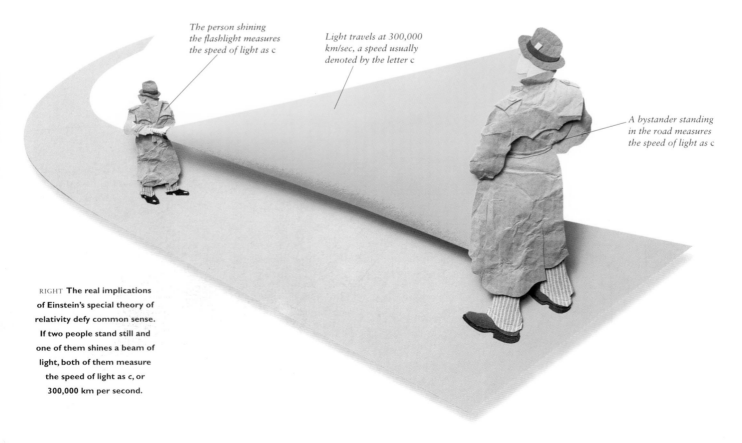

The person shining the flashlight measures the speed of light as c

Light travels at 300,000 km/sec, a speed usually denoted by the letter c

A bystander standing in the road measures the speed of light as c

RIGHT **The real implications of Einstein's special theory of relativity defy common sense. If two people stand still and one of them shines a beam of light, both of them measure the speed of light as c, or 300,000 km per second.**

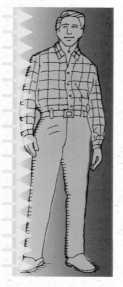

*The speed of
light is fixed*

ABOVE **If you shine a light
from a speeding car, the
light beam travels past a
bystander at the normal
speed of light. The speed of
the car makes no difference.**

the car is *c*, relative to the road it is *c* + 60, and relative to you it is *c* + 66. But Maxwell's equations do not allow for this; they say that the speed of light is *c*, period. The equations take no account of where you are measuring from, or whether the source of the light is moving relative to you. The speed is always *c*, or 300,000 km per second, and no more. (Strictly speaking, the speed of light *in a vacuum* is always *c*; the speed is reduced when the light is travelling through air, or glass, or any other medium.)

Because of this, Maxwell's equations contravened Newton's laws. At the end of the nineteenth century Newton's laws had been around for more than 200 years, and had passed every test applied to them; Newton's position as the greatest scientist who ever lived was unquestioned. Maxwell's equations had been around for three decades, and Maxwell, although highly respected, was not perceived in the same light as Newton. It was natural for scientists to regard the disagreement between Maxwell's

equations and Newton's laws as indicating that something was wrong with Maxwell's equations – especially since Newton's laws agree with common sense.

A FRESH LOOK AT NEWTON

Einstein's great insight was to accept Maxwell's equations at face value, and look at how Newton's laws would have to be modified to be compatible with Maxwell's equations. What would the world be like if the speed of light really is the same number, *c*, for all observers, no matter how they are moving relative to one another or relative to the source of the light? It was this insight that led to the special theory of relativity, published in 1905. This theory unites space and time in one mathematical package, reminiscent of the way Maxwell's equations unite electricity and magnetism. And this should be no surprise since, after all, velocities deal with space and time. Kilometres per second, miles per hour, whatever units you choose – they all involve space and time.

RIGHT **If the person who
shines a beam of light is riding
in a car, both the bystander and
the person with the flashlight
still measure the speed of the
light as c. Every speed in the
universe is relative except for
the speed of light which is
always the same.**

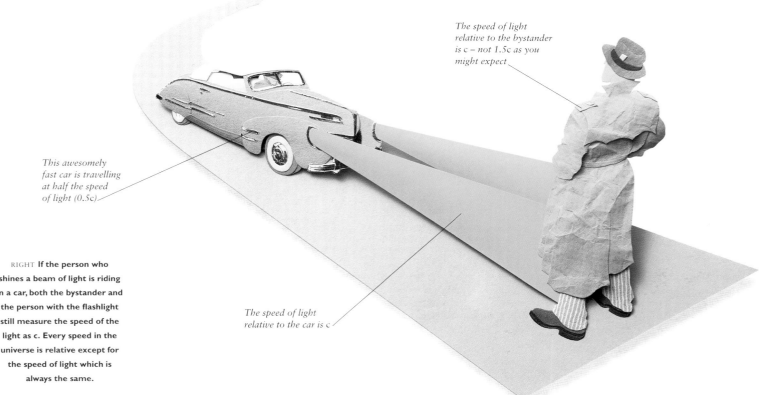

*The speed of light
relative to the bystander
is c – not 1.5c as you
might expect*

*This awesomely
fast car is travelling
at half the speed
of light (0.5c)*

*The speed of light
relative to the car is c*

THE SUPREME LAW

The most important single law in science is not the law of gravity, or any of Newton's laws, or the laws of quantum physics or relativity theory. It goes by the prosaic name of the second law of thermodynamics. If it seems strange that such a fundamental truth should be the second law of anything, that is because the first law of thermodynamics is scarcely more than a preamble, telling us that heat is a form of energy, work and heat are interchangeable, but the total amount of energy in a closed system always stays the same. If, for example, an ice cube is dropped into a Thermos flask of hot water and the top of the flask is sealed, the ice cube gets hotter (gains energy from the water) and the water gets cooler (it loses energy to the ice) but the total amount of energy in the flask stays the same.

The melting ice cube example is particularly appropriate, because it helps to demonstrate many features of the second law of thermodynamics. First, heat always flows from a hotter object to a cooler object, never the other way around, unless work is done (that is, energy is put in to the system) to make it go the other way. A flask of warm water will not spontaneously form itself into a mixture of hot water and ice cubes, even though this would not involve any change in energy overall.

A domestic refrigerator is a good example of the way heat flows in one direction. The inside of a fridge gets cold because gas expands in the operating part: the sealed pipes and flasks in the back. This has a natural cooling effect. Heat from the fridge flows into the cool gas in the pipes, making the gas hotter and the air in the food compartment cooler. But once the gas has done its job, it is led away around the back to the outside of the fridge, where energy (usually from electricity) squeezes it back into a compressed or even liquid state. This makes it hot, which is why the tubes at the back of a fridge are warm to the touch. There, the excess heat is lost to the outside air before the whole cycle repeats.

The amount of heat generated in this way is always more than the amount of heat removed from the food compartment inside the fridge. So if you set a fridge running with the door open inside a sealed room, the air in the room will get hotter, not cooler. The cooling effect on the air inside the fridge is more than offset by the heat escaping from at the tubes at the back.

THERMODYNAMICS

In the nineteenth century, physicists were encouraged to study this flow of heat from hot objects to cooler objects in great detail, because of the importance of understanding such behaviour – thermodynamics – in the steam engines that powered the Industrial Revolution. Thermodynamics was big science in those days, in the same commercially important sense that microchip technology (and therefore quantum physics) has become big science in the last quarter of the twentieth century. These thermodynamic pioneers realized that the flow of heat is linked to another feature of the everyday world: the tendency for disorder to increase – or, in common language, for things to wear out. This is another manifestation of the second law of thermodynamics.

Again, the ice cube analogy helps. When you drop the ice cube into the hot water, there is a certain amount of order – a pattern – in the system. There is a distinction between the ice and the water. But once the ice has melted, you are left with a uniform, featureless flask of warm water. The pattern has gone. This encapsulates the way things happen all around us. If a car is left standing, it will rust and fall apart. No energy input is required. It just happens, relentlessly. But you can't acquire a new car by piling up the bits needed and waiting for nature to assemble them. The only way to

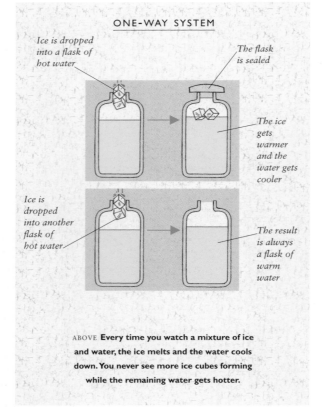

ONE-WAY SYSTEM

Ice is dropped into a flask of hot water

The flask is sealed

The ice gets warmer and the water gets cooler

Ice is dropped into another flask of hot water

The result is always a flask of warm water

ABOVE Every time you watch a mixture of ice and water, the ice melts and the water cools down. You never see more ice cubes forming while the remaining water gets hotter.

get the car assembled is to do some work (put in energy) to put the parts together. What's more, the process of making something structured and orderly like a car always involves making more mess somewhere else – such as in the mines where the metals that go in to the vehicle are mined.

THE COMING OF CHAOS

The same thing is true in nature. It may seem that life violates this law, because living things take unstructured material (water, carbon dioxide and so on) and turn it into elaborate structures (plants and animals). But all of this requires an input of energy, and ultimately the source of this energy is the Sun. Physicists measure order (or rather, disorder) in terms of a quantity known as entropy, which defines just how ordered a system is. Another way to express the second law of thermodynamics is to say that in a closed system, or in the whole Universe, entropy always increases. But the Earth is not a closed system. It is bathed in energy from the Sun. Any decrease in entropy caused by the activity of life on Earth is vastly more than compensated for by the increase in entropy going on as the hot Sun pours energy out into the cold Universe, in line with the second law.

The law is so fundamental that in the 1920s the pioneering astrophysicist Arthur Eddington summed it up in these words,

The law that entropy always increases – the second law of thermodynamics – holds, I think, the supreme position among the laws of Nature. If someone points out to you that your pet theory of the Universe is in disagreement with Maxwell's equations – then so much the worse for Maxwell's equations. If it is found to be contradicted by observation – well, these experimentalists do bungle things sometimes. But if it is found to be against the second law of thermodynamics I can give you no hope; there is nothing for it but to collapse in deepest humiliation.

Part of the reason why Eddington (and others) felt so strongly about the second law (and still do) is that it is bound up with our understanding of the arrow of time, and has literally universal

ABOVE Living things seem to violate the second law of thermodynamics. A growing plant takes in water and nutrients from the soil and carbon dioxide from the air, and creates a complicated, ordered structure – the adult plant. But it can only do this because it is using energy from sunlight. The amount of order created by the growing plant is always less than the amount of order destroyed inside the Sun in making the sunlight that is used by the plant.

implications. Going back to the ice cube in the flask, there is no doubt about which state of the flask comes earlier in time, even though it always has the same energy. The state with hot water and ice is earlier in time; the state with warm water and no ice is later in time. And if we look at what is happening at the level of atoms and molecules, the supreme authority of the second law even over Newton's laws becomes clear.

Imagine a box with a partition dividing it in half. On one side of the partition there is gas, atoms bouncing around in accordance with Newton's laws. On the other side, there is a vacuum. If you take the partition away, the gas spreads out to fill the whole box, cooling down as it does so. But if you sit and watch the box forever, you will never see the gas gather in one half of the box, so that you can drop the partition back in and trap the gas there. The state of the box with gas confined in one end lies in the past, and the state with gas filling the whole box lies in the future, relatively speaking.

THE UNIVERSE IN REVERSE

This is true even though, according to Newton's laws, every collision between the atoms in the gas is precisely reversible. If you had a super-microscope and could film a collision between atoms, you could run the film backwards and what you saw would make perfect sense. But if you made a film of the box as a whole, with gas spreading out to fill the box, running the film backwards would make no sense at all.

The second law defines the arrow of time – the future is the direction in which entropy increases. The future is also the direction in which the Universe, which has been expanding ever since the Big Bang, is bigger. Some physicists have puzzled over whether there is a connection between this cosmic arrow of time and the thermodynamic arrow of time.

If the Universe were contracting, would time run backwards, with ice cubes forming spontaneously in puddles of water? Nobody knows, and, since the Universe is likely to carry on expanding for at least another hundred billion years, nobody need lose too much sleep over it.

SPECIAL RELATIVITY

Einstein found that in order for the speed of light to be measured as the same number, c, for all observers, velocities cannot add up in the way Newton's laws say they do. It turns out that nothing can ever travel faster than the speed of light, and that there is a law of diminishing returns which means that adding two velocities together has a smaller and smaller effect for velocities closer and closer to the speed of light. At the speed of light, adding more velocity has no effect at all, which is why the speed of the light from my car is the same for me in the car as it is for you by the roadside. It isn't just that the extra speed of the car is too small to notice; it is literally true that $c + 60 = c$.

The theory also predicts that an object that is moving relative to an observer will be heavier than when it is at rest, and will shrink in the direction

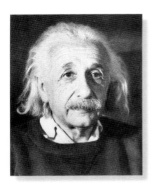

ALBERT EINSTEIN
1879–1955

Best remembered for his theories of relativity, Einstein was also one of the founding fathers of quantum physics, and was the first person in modern times to suggest that light might exist in the form of particles. It was for this work, not relativity, that he received the Nobel Prize (he should, of course, have had two).

of its motion, while a moving clock will run slower than an identical clock that is stationary relative to the observer. And it is the special theory of relativity which gives us the most famous equation in science, $E = mc^2$. All of these effects (except $E = mc^2$!) are tiny unless you are travelling with velocities close to the speed of light, which is why they are not part of everyday experience and common sense.

Yet they have all been tested to very great accuracy in experiments, and the results of those experiments are precisely in line with the predictions of the theory of relativity. Einstein's model of the world is a very good scientific model, and for any velocities that are much less than c it gives exactly the same answers as Newton's laws, as you would expect. But unlike Newton's laws, it is fully consistent with Maxwell's equations.

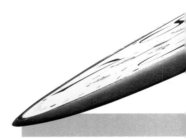

ELASTIC TIME

The best way to get a picture of what the special theory of relativity is all about is to take up the image of spacetime as a four-dimensional continuum – an idea first proposed by Hermann Minkowski in 1908.

This geometrization of the special theory of relativity (which, it should be said, is a fully worked-out mathematical treatment, although we give just the physical analogies here) made it accessible to other scientists, and established Einstein's early reputation. This is ironic, since Minkowski had been one of Einstein's tutors when he was an undergraduate at university, and described Albert at that time as a 'lazy dog' who 'never bothered about maths at all'.

The feature of relativity theory that flies in the face of common sense and makes it hard to relate to everyday experience is the way objects shrink when they move, and moving clocks run slow, so that time stretches out for them. The warping of spacetime

that leads to the occurrence of black holes is not so hard to understand because Newtonian gravity, which we all think we have a feel for, also predicts the existence of objects with a gravitational pull so strong that not even light can escape from them. But time dilation and space contraction are really weird ideas, at first sight, with no obvious counterparts in the conventional Newtonian model.

Minkowski's geometrical picture of spacetime gives an insight into what is going on. In the three dimensions of everyday space, an object has a distinct size and shape. But depending on the angle from which you view it, you may see the object differently.

To keep things simple, consider just a straight rod with a certain length. If you look at the rod directly from one side, its full length will be apparent. If you look at it end-on, it will look like a small disc. At intermediate angles it will appear foreshortened – we even use a term that includes the word 'shortened' to describe its

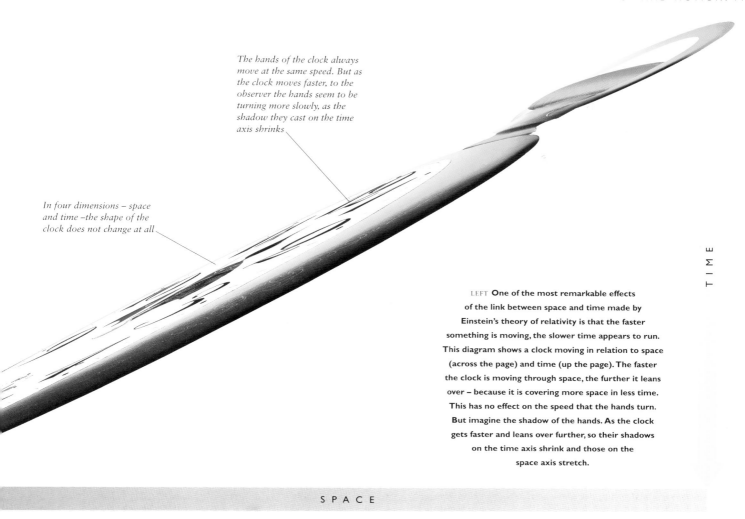

The hands of the clock always move at the same speed. But as the clock moves faster, to the observer the hands seem to be turning more slowly, as the shadow they cast on the time axis shrinks

In four dimensions – space and time –the shape of the clock does not change at all

TIME

LEFT **One of the most remarkable effects of the link between space and time made by Einstein's theory of relativity is that the faster something is moving, the slower time appears to run. This diagram shows a clock moving in relation to space (across the page) and time (up the page). The faster the clock is moving through space, the further it leans over – because it is covering more space in less time. This has no effect on the speed that the hands turn. But imagine the shadow of the hands. As the clock gets faster and leans over further, so their shadows on the time axis shrink and those on the space axis stretch.**

SPACE

appearance. Yet the rod has the same 'real' length. Now consider the dimensions of space and time represented by two lines at right angles to one another, like the axes of a graph. Time is usually represented 'up the page', and space 'across the page', making a kind of graph called a spacetime diagram. In the four dimensions of real spacetime (or the two dimensions of this model of spacetime), an object has a certain size, which is called its extension. This is a property involving both space and time that is equivalent to the length of the rod in ordinary, three-dimensional space.

You can represent this spacetime extension of an object by placing a matchstick on the spacetime diagram. The matchstick always has the same length. But depending on how the matchstick is rotated on the diagram, its length measured along either of the two axes will vary. As you rotate the matchstick, its 'shadow' on the time axis stretches while its shadow on the space axis shrinks.

Movement through space is equivalent to making this kind of rotation in the four dimensions of spacetime. The space part of the extension gets smaller, the time part gets bigger in proportion, and the overall extension stays the same.

When an object slows down, its extension rotates back the other way, shrinking the time component and stretching the space component back towards the values they have when the object is stationary. But still the extension – that is its total length in four dimensions – stays the same.

If relativity theory were to predict that moving objects shrink but time stays the same for them, or that moving clocks run slow but their shape stays the same, we would say that this is really weird, and also incomprehensible. However, because the two properties change in the opposite sense but by equivalent amounts, relativity theory makes perfect sense – in four dimensions.

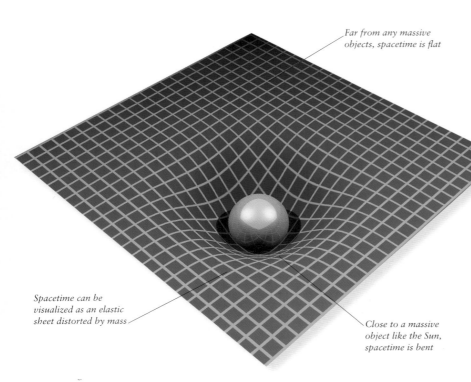

Far from any massive objects, spacetime is flat

Spacetime can be visualized as an elastic sheet distorted by mass

Close to a massive object like the Sun, spacetime is bent

GRAVITY AND SPACETIME

The special theory of relativity is 'special' because it is a restricted theory, a special case of a proper full theory of how objects move. This is because it deals only with things moving at constant velocities. Just over ten years later, Einstein succeeded in developing a general theory of relativity, which includes all of the special theory of relativity and also deals with the behaviour of accelerated objects – objects whose speed is changing. Even better, the general theory also explains the nature of gravity, using an extension of Faraday's idea of a field.

ABOVE **Spacetime is like the web of forces imagined by Michael Faraday. The forces are straight if there are no gravitational forces operating, but any massive object bends the lines of force. This deflects the motion of any objects travelling through spacetime into curved trajectories.**

THE COSMIC TRAMPOLINE

The general theory of relativity involves some horribly complicated equations (actually, the equations are rather elegant, mathematically speaking; it is solving them that is horribly complicated) but happily it can be understood in physical terms by regarding spacetime as a four-dimensional continuum.

Because people are not good at visualizing four dimensions, the usual trick is to think of a stretched rubber sheet as representing spacetime in two dimensions, with one direction across the sheet corresponding to space and the direction at right angles representing time.

If there are no gravitational fields spacetime is flat, like the stretched surface of a trampoline. But the presence of an object like the Sun is equivalent to placing a heavy object such as a bowling ball on the trampoline, making a dent in it. So spacetime is curved by the presence of matter, and the more massive the matter, the more spacetime is curved.

In flat spacetime, objects move in straight lines, like a marble rolled across the flat surface of the trampoline. But in curved spacetime, objects are deflected as they follow the curvature. Gravity is seen as operating by making dents in spacetime, so that any objects moving through curved spacetime follow curved trajectories. In a nutshell, matter tells spacetime how to bend, and spacetime tells matter how to move.

LEFT **Despite being described by his university tutor as someone who 'never bothered about maths at all', Einstein was a brilliant mathematician whose revolutionary theories were solidly grounded in irrefutable mathematical logic.**

*Every observer, anywhere in the
Universe, seems to be in the middle
of an expanding bubble. We are all
surrounded by the Big Bang*

EXPERIMENTAL PROOF

Einstein's theory predicted that, as a result of this process, light passing close by the edge of the Sun would be deflected by a certain amount. In 1919 an eclipse of the Sun gave astronomers the chance to photograph light from distant stars grazing the edge of the Sun, without the glare of sunlight fogging their photographic plates. When these photographs were compared with photographs of the same part of the sky taken six months earlier, when the Sun was on the opposite side of the Earth, they showed that the light from distant stars was shifted sideways towards the Sun, by exactly the amount Einstein's general theory had predicted.

It was this experiment that made Einstein famous, establishing his image as the greatest scientist since Newton. It also established the general theory of relativity as the best theory of gravity that has ever been proposed. And the greatest achievement of the general theory is its application to the entire Universe, to explain how everything began in the Big Bang, some 15 billion years ago *(see Chapter 3)*.

ABOVE **During a total eclipse of the Sun, the sky is dark enough to see stars that appear next to the Sun in the sky, although really they are much further away.**

BELOW **The popular image of the Big Bang is like an explosion in space. But in the real Big Bang, space itself was created in the explosion. There was nothing 'outside' to explode into.**

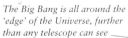

The Big Bang is all around the 'edge' of the Universe, further than any telescope can see

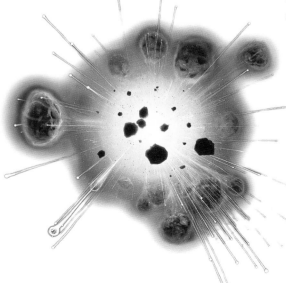

RIGHT **Space and time all began with the Big Bang which started the universe some 15 billion years ago. As we look at the most distant galaxies, some 12 billion light years away, we are looking back into, literally, the dawn of time, 12 billion years ago.**

FISSION AND FUSION

The practical power of the two great theories of twentieth-century physics, relativity theory and quantum theory, is demonstrated by their success in explaining two processes that operate on the scale of the nuclei of atoms, but affect our daily lives. Nuclear fission provides the power of the atomic bomb and the nuclear reactors used to generate electricity; nuclear fusion provides the energy which keeps the Sun shining. These processes release energy because of the way energy is stored in the nucleus of an atom. The process is explained by quantum physics, and the amount of energy that is released is predicted by relativity theory.

NUCLEAR FUSION

Atomic nuclei are made up of protons and neutrons. The simplest atom, hydrogen, has a nucleus consisting of a single proton. The most common form of the next lightest element, helium, has a nucleus containing two protons and two neutrons. In all other cases, stable nuclei contain at least as many neutrons as protons, because neutrons are needed to help hold the nuclei together in spite of the tendency of the positive charge in all the protons to blow it apart.

But although a nucleus of helium contains two protons and two neutrons, its mass is not the same as the mass of two isolated protons and two isolated neutrons added together. When primordial particles combine to make helium nuclei (this happened in the Big Bang, and still happens inside stars like the Sun today), some of the mass involved (0.7 per cent of the mass of the helium nucleus) is turned into energy, chiefly in the form of electromagnetic radiation. This process is nuclear fusion, and it happens precisely because the heavier nucleus (in this case, helium) stores less energy than the lighter nucleus, so it is more stable.

The amount of energy E released by the conversion of mass m is given by Einstein's famous equation $E = mc^2$, where c is the speed of light. As everyone knows, c is a huge number – 300,000 km per second – so you get an awful lot of E for a small amount of m. The mass difference between hydrogen and helium nuclei can be measured, and the energy released can also be measured, confirming the accuracy of this equation. It is this process of fusion of hydrogen into helium (through some intermediate steps which convert some protons into neutrons) that keeps the Sun shining.

FORGING IRON

Up to a point, the more massive nuclei of heavier elements are even more efficient in terms of energy storage than helium; that is, they need proportionally less energy to bind themselves together. Protons and neutrons are collectively called nucleons, and the energy stored per nucleon gets progressively less all the way up to iron-56, which has 56 nucleons (26 protons and 30 neutrons) in each atomic nucleus. The iron nucleus provides the most energy-efficient way to pack nucleons together, and because of this the successive stages of nuclear fusion inside stars build all the elements from hydrogen and helium to iron.

ABOVE **Our knowledge of sub-atomic particles was boosted dramatically by American physicist Don Glaser's invention of the bubble chamber in 1952. When electrically charged particles shoot through liquid hydrogen in the chamber, they leave a trail of bubbles in their wake which marks their path clearly on a photograph. The magnetic field around the chamber curves positively charged protons one way and negatively charged electrons the other. By studying bubble chamber tracks, physicists have learned a great deal about the way subatomic particles interact.**

ABOVE **The energy of an exploding hydrogen bomb is released by nuclear fusion. The reaction is triggered by a nuclear fission device, which recreates the extreme conditions that exist inside the Sun.**

HEAVY ELEMENTS

Nuclei with more nucleons than iron-56 store the nucleons less efficiently than iron does. This is essentially because the mutual repulsion of all the individual protons is starting to become a real problem for those nuclei. In order to make such nuclei, energy has to be put in, turning some of the E back into m. This happens in the explosive death throes of stars called supernovas, building up heavier elements such as tin, gold, lead and uranium, and scattering them into space along with the mixture of elements built up earlier by the star. It is this mix of stardust that eventually makes planets like the Earth, and people like us.

The overall effect of the way the amount of energy per nucleon differs from one nucleus to the next is as if the different elements sit on little ledges up the sides of a valley of stability, with iron at the bottom of the valley. Lighter elements, all the way up to hydrogen, go up one side of the valley; heavier elements go up the other side of the valley. Under the right conditions, the lighter elements can be given a nudge which encourages them to move down the ledges towards the valley floor, building up bigger nuclei by fusion. On the other side of the valley, heavier elements can be encouraged to move down the ledges towards the valley floor by the reverse process, fission, splitting the nuclei apart to make lighter ones. In some cases, the heavy nuclei are unstable and decay spontaneously. In other cases, they do so if they are hit by a stray neutron or other particle.

NUCLEAR FISSION

Fission is the process used in all successful nuclear reactors to date, and in the atomic bomb (really a nuclear bomb). The energy released by fission today is the energy that was put in to make these heavy elements long ago; it is literally the stored energy of a supernova. In a bomb, the energy stored in an unstable heavy element such as uranium-235 (each nucleus containing 92 protons and 143 neutrons) is released explosively once the reaction is triggered. This is because each nucleus that splits releases several neutrons, which each

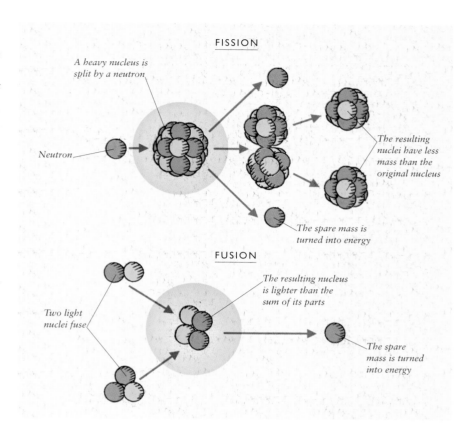

FISSION

A heavy nucleus is split by a neutron

Neutron

The resulting nuclei have less mass than the original nucleus

The spare mass is turned into energy

FUSION

The resulting nucleus is lighter than the sum of its parts

Two light nuclei fuse

The spare mass is turned into energy

trigger the splitting of more nuclei. In a power reactor the process is controlled by using material which captures most of the neutrons, so that on average each nucleus that splits triggers the fission of just one more nucleus. Both processes are examples of chain reactions.

Fusion is the process which may one day provide an almost unlimited source of energy on Earth, tapping the same energy supply as the stars. The problem is that in order to make positively charged nuclei (such as those of hydrogen) fuse together, they have to be squeezed together with sufficient force to overcome their electrical repulsion (the very thing that makes fission of heavy elements relatively easy) and allow the stronger nuclear forces (essentially, the colour force) to take over. This means recreating the conditions that exist inside a star, in reactors here on Earth. No wonder, then, that some people think that it makes more sense to try to tap in to the energy supply from nature's very own tried and tested fusion reactor, the Sun itself.

ABOVE **Nuclear fission and nuclear fusion both result in a net loss of nuclear mass. The missing mass is transformed into energy according to Einstein's equation, in which the energy equals the mass multiplied by the speed of light, squared.**

WARM

HOT

RED HOT

WHITE HOT

ABOVE **When an object gets hotter, it glows first red hot then white hot. But classical physics says this is impossible – it should always glow the same colour. Max Planck's brilliant insight was to explain the colour of light in terms of little packets or 'quanta' of energy.**

QUANTUM THEORY

By the time the general theory of relativity had supplanted Newton's theory as the best theory of gravity we have, most physicists were more concerned about what was going on at the other end of the scale of sizes, in the world of the very small. As they had begun to probe the structure of atoms at the end of the nineteenth century, they had found more examples of behaviour which did not fit in with Newton's laws. This led to the other great theory of twentieth-century physics, the quantum theory. Quantum theory is one of the most astonishing scientific breakthroughs of all time and completely upsets our notions of cause and effect. Once again the theory began with a puzzle about the nature of light.

PARTICLES OF LIGHT

The first step was taken by the German physicist Max Planck, right at the end of the nineteenth century. Like many of his colleagues, he was puzzled by the way hot objects radiate light. We all know that a warm object radiates heat but does not glow, yet a hotter object glows red hot, an even hotter object glows white hot, and so on. In terms of wavelengths, the hotter objects radiate more light with shorter wavelengths. But classical theory (which means anything pre-quantum; even the general theory of relativity is a classical theory) says that all the energy in a hot object should be released in a burst of short-wavelength light, whatever its temperature.

Planck found that he could explain the nature of radiation from a hot object (called black body radiation, because it is the radiation you get if a perfectly black object is heated up) if he assumed that the atoms inside the object radiate light only in packets of a certain size, which he called quanta (a single packet is called a quantum). Short-wavelength light carries a lot of energy in each packet, and individual atoms don't have enough energy to make quanta that big at low temperatures. This explains why they cannot, after all, release all their energy in the way classical theory predicts.

But Planck did not suggest that light exists only as quanta, little particles of light. He thought that the reason why black bodies radiate in this way had to do with the nature of atoms, not the nature of light. It was Albert Einstein who, once again, made a complete break with tradition – in 1905, the same year in which he published his remarkable special theory of relativity.

MAX PLANCK
1858–1947
The physicist who first used the idea of light quanta to explain the nature of the electromagnetic radiation from hot objects. Rather neatly, his ground-breaking paper was published in 1900, just at the start of the twentieth century.

BELOW **Choreographers may not always like it, but individual human beings have free will. We are free to make mistakes, or to follow our own whims about what to wear. Some physicists believe that free will can be traced back to the uncertainty built into the laws of quantum mechanics.**

The choreographer wants all the girls in the chorus line to look the same

She may not be cut out for the chorus line, but this dancer knows what colour she likes her hair

Some people don't wear what they are told. They exercise free will

All the girls are supposed to wear identical outfits

UNCERTAINTY

One of the most important features of the quantum world is also one of the most misunderstood. In the 1920s, Werner Heisenberg formulated his principle of uncertainty, which says that a quantum entity (such as an electron) does not have both a precise position and a precise velocity at the same time. At any given moment, the electron itself does not 'know' both exactly where it is and exactly where it is going. This is linked to the idea of wave-particle duality – obviously, a wave does not exist 'at a point' in the way a particle can.

This uncertainty is intrinsic to the nature of quantum reality, and in fact all of quantum mechanics can be constructed from Heisenberg's uncertainty principle. Unfortunately many textbooks, even at university level, say that the uncertainty is a result of our human limitations, and the fact that our experiments cannot make measurements that are sufficiently precise. This is not true! Uncertainty is a real feature of the quantum world, and this is what makes the quantum world run in accordance with the rules of probability.

You may not like this, but you should. If the world ran in strict accordance with Newton's laws, like clockwork, then everything, down to the tiniest particle interaction, would be determined in advance. There would be no scope for free will. It is quantum uncertainty that gives us back the chance to run our own lives and make our own decisions rather than follow a preordained plan.

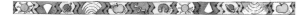

THE PHOTOELECTRIC EFFECT

Einstein had been puzzling for some time over the photoelectric effect, a process in which the energy from a beam of light shining on a metal surface knocks electrons out of the metal. Many experiments had shown that for light of a certain colour – which means a precise wavelength – all the electrons knocked out in this way have the same energy, regardless of the strength of the beam.

You might think that if the beam of light were made weaker, the electrons would have less energy. But they don't. If the beam of light is weaker, fewer electrons get knocked out of the metal, but those few have the same energy as the electrons produced when the beam of light is stronger. Yet the energy of electrons knocked out by blue light is always bigger than the energy of electrons knocked out by red light. Shorter-wavelength light carries more energy in each punch.

Einstein's explanation was that light really does exist in the form of little particles, now known as photons. All the photons in blue light have a certain energy. An electron is knocked out of an atom by the impact of a single photon, so it always carries that amount of energy. If the beam is weak, there are simply fewer photons around. The same is true for red light, except that each photon of red light has less energy than a photon of blue light.

Einstein's suggestion did not meet with approval in 1905. It went against a century or more of apparently convincing evidence that light travels in waves, as originally suggested by Huygens, not in particles, as Newton had suggested at roughly the same time. Robert Millikan, an American physicist, was so annoyed by Einstein's idea that he spent ten years trying to prove that Einstein was wrong by carrying out a series of increasingly accurate experiments. He succeeded only in proving that Einstein was astonishingly absolutely right, and that photons – little particles of light – really do exist. It was as a result of that work, finally completed in 1915, that Einstein received his Nobel Prize in 1922; Millikan received the Prize in 1923.

THE TELL-TALE LINES

The nature of the light radiated (or absorbed) by atoms depends on the rules of quantum physics. The study of the light radiated by atoms – spectroscopy – helped to provide an understanding of those quantum rules, and an understanding of the quantum rules helps physicists to build mathematical models of how atoms work. And because each kind of atom (each element) has its own unique spectral fingerprint, physicists are able to use spectroscopy to probe the composition of material, both in the laboratory and in stars.

The way light is emitted or absorbed by an atom depends on the way electrons are arranged in the outer part of the atom. This is complex, but for our purposes we can use the simple model developed by Niels Bohr, in which the electrons are seen as little particles 'in orbit' around the nucleus of the atom.

QUANTUM LEAPS

When an atom emits or absorbs light, it does so only in packets of energy with a definite size – photons. Absorbing a photon provides the energy for an electron to move from an orbit close to the nucleus to an orbit further out from the nucleus. When and if the electron falls back into its old orbit, it emits a photon with the same energy: the energy corresponding to the gap between the orbits.

The important thing to grasp is that there is no in-between state, because the electron cannot emit or absorb a fraction of a photon. First it is in one orbit (or one energy level), then it is in another, instantaneously. The other important point is that the energy of a photon is precisely related to its

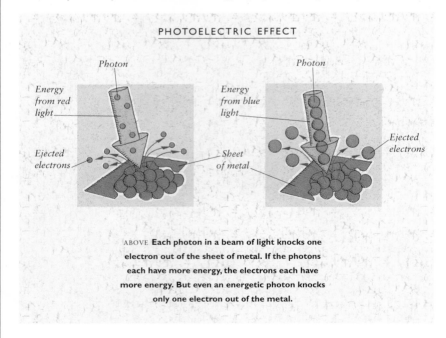

PHOTOELECTRIC EFFECT

Photon

Energy from red light

Ejected electrons

Photon

Energy from blue light

Sheet of metal

Ejected electrons

ABOVE Each photon in a beam of light knocks one electron out of the sheet of metal. If the photons each have more energy, the electrons each have more energy. But even an energetic photon knocks only one electron out of the metal.

ELECTROMAGNETIC WAVES

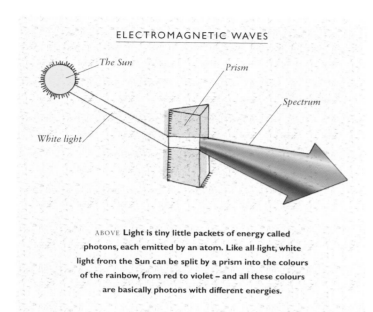

ABOVE Light is tiny little packets of energy called
photons, each emitted by an atom. Like all light, white
light from the Sun can be split by a prism into the colours
of the rainbow, from red to violet – and all these colours
are basically photons with different energies.

wavelength, so a photon with a precisely determined energy
corresponds to light with a precisely defined wavelength, or colour.

The whole rainbow spectrum of visible light runs from red
through orange, yellow, green, blue and indigo to violet. Indeed, the
spectrum extends beyond these visible limits to both longer
wavelengths (beyond red) and shorter
wavelengths (beyond violet).

Each different wavelength in the
spectrum corresponds to a particular
energy emitted by an atom as
electrons jump from orbit to orbit,
falling back to a lower energy level.

If an atom emits light at a precise
wavelength, this produces a bright,
sharply defined line in the spectrum. If
it absorbs light, there is a dark line.
Because the arrangement of electrons
in the atoms is different in every
element, the pattern of lines made by
the allowed electron jumping is
different for each element. Which is
why astronomers can look at the light
from a distant star, split it into its
rainbow spectrum using a prism, and
find out what the star is made of.

ROBERT MILLIKAN
1868–1953
Millikan made extremely
accurate measurements of
the size of the charge on the
electron, proved Einstein's
photon theory of light was
correct (even though he
didn't want to) and
determined the value of
Planck's constant.

FRAUNHOFER'S SPECTRAL LINES

The discoveries that led to spectroscopy becoming such a powerful
tool were made by the German physicist Josef von Fraunhofer in
1814. He was actually interested in the properties of glass, and
passed light through prisms in order to study the glass. He used
light from the Sun, which had previously been thought of as 'pure'
and unsullied, but despite this he noticed many dark lines in the
spectrum produced by each prism.

At the end of the 1860s, studies of the spectrum of sunlight
showed (along with traces of many known elements) a set of lines
that corresponded to no element known on Earth. The British
astronomer Norman Lockyer said that they must belong to
a 'new' element, which he dubbed helium, from the Greek word for
the Sun. Helium was eventually identified on Earth in 1895,
and found to have exactly the properties required to explain the
solar lines.

ABOVE We get colours because each kind of atom emits photons
at a particular energy. White light is a mixture of a wide range of photons
of different energies. But by sending white light through a prism, all the
colours it contains can be separated out into a spectrum, a rainbow range
of colours. Because colour acts like a kind of identity tag for the atoms in
the original light source, we can tell what kind of atoms make up the
light source – whether it is an electric light or a distant star. This is
the basis of the science of spectroscopy.

THE WAVE-PARTICLE DUALITY

By the time Einstein received the Nobel Prize it was clear that although light may travel like a wave, it interacts – or arrives – like a particle. It has the properties of both waves and particles. What's more, experiments involving electrons soon showed that, although they had previously been regarded as particles, they interfere with one another in experiments equivalent to Young's double-slit experiment with light. So electrons also have both particle-like and wave-like properties in one package. By the end of the 1920s this idea of wave-particle duality had become the foundation of a complete theory of the subatomic and atomic world, a theory known as quantum physics.

The chief architect of the theory, in the form in which it has usually been taught ever since, was the Danish physicist Niels Bohr. The theory is often referred to as the 'Copenhagen Interpretation' of quantum physics in Bohr's honour, although he drew on the work of the Austrian Erwin Schrödinger, the Germans Werner Heisenberg and Max Born, and the Englishman Paul Dirac. It certainly highlights the oddities of the quantum world, and you may find it hard to accept, but like the general theory of relativity it has passed every experimental test.

DISSOLVING PARTICLES

The Copenhagen Interpretation says that quantum entities, such as photons and electrons, interact in accordance with the rules of probability, and only exist in the form of what we would think of as real particles at the moment they are being observed.

The simplest way to picture this is to imagine either a photon or an electron confronted by the familiar double-slit experiment. In the everyday world, you would expect the 'particle' to go through one slit or the other. But according to the Copenhagen Interpretation, a 'wave of probability' associated with the quantum entity goes through both slits at once, and interferes with itself. It is the result of that interference that decides where the 'particle' itself will pop up on

NIELS BOHR
1885–1962

A great pragmatist, Bohr didn't worry too much about what quantum theory meant as long as he could find rules which made it work – in the sense of predicting things that could be tested by experiments. He developed the first successful quantum model of the atom.

the other side of the screen. It is as if the particle dissolves into a wave (known as the 'wave function'), and then the wave collapses back into a particle when you try to measure its position. But as soon as you stop looking at it, the particle starts spreading out like a wave again.

The idea that the world is ruled by probabilities led Einstein to break with the quantum theory he had helped found, saying, 'I cannot believe that God plays dice'. But every experiment has shown that probability does indeed rule in the quantum world. Happily, though, just as the theory of relativity reduces to Newton's laws at low speeds, so quantum theory reduces to Newton's laws for anything much bigger than an atom. Which is why things like billiard balls are solid objects that stay in one place, and don't dissolve into probability waves if you turn your back on them.

Many physicists find this whole package of ideas abhorrent, and have sought for decades for a way round it. As yet, they have not succeeded, although there are various interpretations of quantum physics which give exactly the same answers as the Copenhagen Interpretation.

ABOVE The quantum leap is an instantaneous transition from one state to another state. There is no 'in between.' If a rabbit moved like a quantum entity, it would disappear in one place and reappear somewhere else, instantaneously, without taking any time to cross the gap.

PAUL DIRAC
1902–84

The quiet man of quantum physics, Dirac proved that the particle approach of Werner Heisenberg and the wave approach of Erwin Schrödinger were mathematically equivalent to one another. The quantum world is particle and wave at the same time.

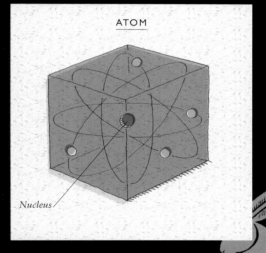

ABOVE **Electrons in atoms behave like quantum rabbits. An electron in one 'orbit' can disappear, and reappear in another 'orbit'. When it does so, a photon with exactly the energy corresponding to the difference between the two 'orbits' is either absorbed or emitted, making a line in the spectrum.**

familiar image of the atom: a tiny central nucleus, carrying a positive electric charge, surrounded by even tinier electrons (each negatively charged) in some sense 'in orbit' around the nucleus. In fact, you need quantum physics to explain how the electrons form a cloud around the nucleus, and why they do not fall into the nucleus. The largest atom is just 0.0000005mm across, but the size of the nucleus in proportion to the size of the electron cloud is roughly the same as that of a golf ball in the centre of a football stadium.

ELECTRONS, PROTONS AND NEUTRONS

The nuclei of atoms contain both positively charged particles, called protons, and neutral particles called neutrons (except for the nucleus of the lightest and simplest atom, hydrogen, which contains just one proton). The difference between each element is essentially the number of protons in its atomic nucleii – so for example a platinum atom has 78 protons, but add another and you have gold, with 79.

The number of negatively charged electrons in the cloud around a nucleus is the same as the number of protons in the nucleus, so the electric charges cancel out and atoms are electrically neutral overall. The number of protons determines the number of electrons, and the electrons and their arrangement in the cloud determine the way the atom interacts with other atoms – its chemical properties. But it is the nucleus that determines the mass of an atom. Protons and neutrons have about the same mass as each other, each nearly 2000 times heavier than an electron.

Rutherford had originally probed the structure of atoms by firing 'alpha particles' at them. Alpha particles are produced when unstable (radioactive) atoms spit them out from their nuclei. When Rutherford's team fired alpha particles at a thin sheet of gold foil, they found that while most of the particles passed right through without being affected by the gold atoms, some hit something solid and bounced back. It was these studies that showed that atoms are mostly empty space, with the mass concentrated in a tiny central nucleus.

INTO THE ATOM

The greatest triumph of quantum physics came in the 1940s, when it was applied to explain the way light (and all electromagnetic radiation) interacts with charged particles. This theory of quantum electrodynamics, or QED, explains the interaction between matter and light in quantum terms, just as Maxwell's equations explained everything about light in classical terms. Even better, physicists found that it could be used to describe how particles interacted on scales much smaller than the atom.

The structure of the atom itself had been discovered early in the twentieth century, largely thanks to the experimental work of the New Zealander Ernest Rutherford (by then based in England) and his colleagues. It was Niels Bohr, in the second decade of the century, who interpreted the experimental evidence to come up with the

QUARKS AND GLUONS

By the end of the 1960s, physicists were able to probe the structure within protons and neutrons in the same sort of way, by firing beams of electrons at them and studying the ricochets. (It is an indication of just how small protons and neutrons are compared to atoms that it took 60 years to develop the technique down to this level.) It was these studies that established, in the 1970s, the idea that although electrons are truly fundamental entities with no internal structure, protons and neutrons are each made up of smaller entities, called quarks.

It is now thought that there are three quarks inside each proton, and three quarks inside each neutron. They possess a property which is analogous to the electric charge on protons and electrons, called 'colour charge', or just colour for short. This is only a label, and does not mean that quarks are coloured in the everyday sense of the word. Electrically charged particles are held together (or pushed apart) by exchanging photons, which are the quanta of the electric field. In a similar way, quarks are held together by the exchange of particles called gluons, which are the quanta of the colour field.

The way this happens is described by a theory called quantum chromodynamics, or QCD, which is modelled on the hugely successful theory of quantum electrodynamics. Unfortunately, although there are only two kinds of electric charge (positive and negative) and just one kind of photon to carry electromagnetic forces, QCD involves three different colour charges and requires eight different gluons, which makes the theory much harder to deal with. Even so, it has had many successes.

Among other things, it explains why the positively charged protons in a nucleus don't repel each other and blow the nucleus apart. The attraction between quarks is very short-range, because gluons cannot travel far, and it mainly operates inside protons and neutrons (which is why they have the size they do). But a little of the attraction leaks out into the neighbouring particles

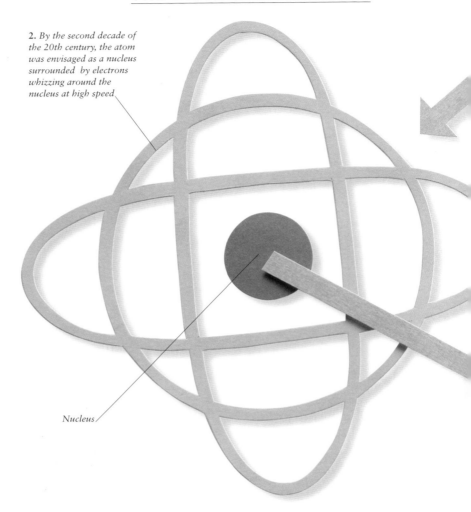

2. By the second decade of the 20th century, the atom was envisaged as a nucleus surrounded by electrons whizzing around the nucleus at high speed

Nucleus

4. The latest discovery is that quarks form each proton and neutron in the nucleus

RIGHT **The atom used to be thought of as a solid, indestructible sphere. Then physicists discovered that it was made up of a tiny central nucleus surrounded by electrons. The nucleus itself was then discovered to be made up of protons and neutrons. We now think that protons and neutrons are made up of quarks. Is this the end of the line?**

Quark

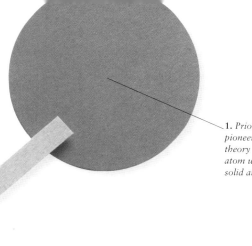

1. *Prior to John Dalton's pioneering work on atomic theory in the 1890s, the atom was envisaged as solid and unbreakable*

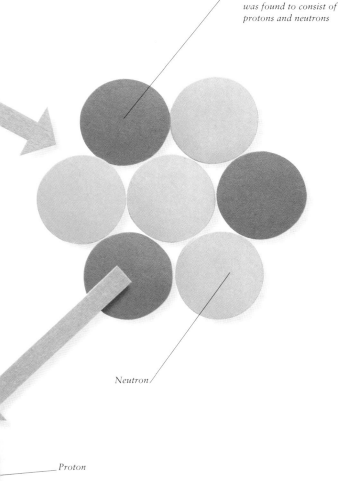

3. *After the discovery of the neutron in the 1930s, the nucleus was found to consist of protons and neutrons*

Neutron

Proton

BELOW **When Rutherford and his team fired alpha particles (essentially helium nuclei) at thin gold foil most of the alpha particles passed straight through, showing that matter is mostly empty space separating heavy atomic nuclei. But some particles hit the gold nuclei and rebounded, showing just how solid a nucleus can be.**

in the nucleus, and over the range of sizes represented by an atomic nucleus this force of attraction – the residue of the glue force – overcomes the electrical repulsion. This explains beautifully, as no other model can, why nuclei have the size they do.

There is one other force which operates on the scale of nuclei, but not in the world at large. It is called the weak nuclear force (because it is weaker than the glue force), and it is responsible for the process of radioactive decay, in which an unstable nucleus spits out an electron and turns itself into a stable nucleus. This force is harder to understand in terms of analogies with the everyday world, but easier to treat mathematically than the colour force because it involves only three photon-like particles of its own. One (like the photon) has zero electric charge, one has positive charge and one has negative charge. Unlike the photon, all three particles have mass.

These entities are so much like photons that it has been possible to develop a single set of equations that describe both electromagnetism and the weak force (much as Maxwell's equations describe both electricity and magnetism). This is the basis of the 'electroweak' theory, which has been tested in many experiments and passed every test. Physicists hope that, one day, it may be possible to combine the electroweak theory and QCD in one mathematical package as a so-called Grand Unified Theory, or GUT.

RUTHERFORD'S ALPHA PARTICLES

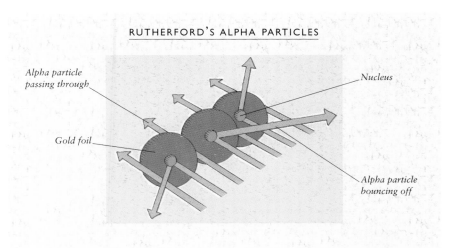

Alpha particle passing through

Nucleus

Gold foil

Alpha particle bouncing off

STRING THEORY

The ultimate goal of physics is to go one step beyond the unification of electroweak theory and QCD, though, and bring gravity into the package. Developing a quantum theory of gravity, even without finding a way to merge it with QCD and electroweak theory, has proved the rock upon which the hopes of many physicists – including Albert Einstein – have foundered. But at the end of the 1990s there is a new air of optimism among the physicists, thanks to a leap of the imagination as bold as the leap Einstein himself took when he rejected Newton's laws in favour of Maxwell's equations. In a move that would have stunned physicists of a century ago, theorists are now developing models which describe the entities we used to think of as particles in terms of tiny loops of vibrating material, prosaically dubbed 'string'.

QUANTUM MUSIC

These loops would be incredibly tiny. It would take a hundred billion billion of them to stretch across a single proton, so there is no hope of probing them directly by experiments. Even so, mathematical models have been constructed which attempt to explain quarks and electrons (and photons and other force carriers) in terms of various different kinds of vibrating strings.

According to these mathematical models, the same kind of string vibrating one way would look like an electron, but vibrating another way would look like a quark. The obvious analogy is with the way different musical notes can be played on a single violin string, each making the string vibrate in different ways.

But in the world of quantum string theory, a strange thing happens. Without any prompting, the equations automatically insist that there must be another kind of string vibration, as well as the ones needed to explain all the familiar particles and quantum forces. At first, this seemed an embarrassing and unwelcome addition, and the physicists tried to get rid of it. But then they realized that this string state corresponds to the properties of a graviton, the quantum of the gravitational field – the equivalent, for gravity, of the photon of light. The ultimate goal – a quantum theory of gravity that could link up with QCD and electroweak theory – seemed to be in sight.

An enormous amount of work still has to be done to find out what this means, and whether gravity, electroweak interactions, colour and all the particles of nature can emerge naturally from a single string theory. But that is what the hot money is riding on as physics moves into the twenty-first century. I wonder what Newton would have made of it.

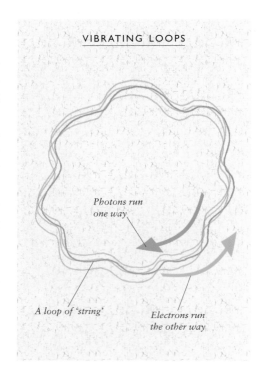

VIBRATING LOOPS

Photons run one way

A loop of 'string'

Electrons run the other way

ABOVE **All the particles ever discovered may, one day, be explained as different vibrations of a single kind of 'string'. In the neatest version of this theory, particles in the everyday sense (things like electrons) are made from waves running one way round a loop of string, while waves in the everyday sense (things like photons) correspond to ripples running the other way around the loop of string.**

LEFT **String theory is very similar to the vibrating strings on a violin – a variety of musical notes can be achieved by a single string vibrating in different ways.**

AT A GLANCE

FALLING BODIES

1638

Galileo Galilei discovers the basic principles of acceleration and falling bodies.

1665-1686

Isaac Newton introduces his theory of gravitation and three laws of motion, and also does pioneering work on the nature of light.

1678

Christiaan Huygens proposes the wave theory of light.

1831

Michael Faraday suggests the idea of magnetic lines of force, and proposes that light can be explained in terms of vibrations of lines of force.

GRAVITATIONAL FORCE

1864

James Clerk Maxwell proposes the idea of fields linking electricity and magnetism, oscillating at the speed of light.

JAMES MAXWELL

1900

Max Planck introduces the idea of quanta.

MAX PLANCK

1905

Albert Einstein proposes his theory of special relativity, in which he demonstrates that the speed of light is constant everywhere in the Universe and everything else is relative.

ALBERT EINSTEIN

1910

Robert Millikan determines the charge on an electron.

1915

Einstein proposes his general theory of relativity about the nature of gravity, in which he predicts how gravity can bend spacetime.

1911

Ernest Rutherford discovers the atomic nucleus (and later discovers the proton).

1913

Niels Bohr develops the Bohr model of the atom with electrons in particular orbits around the nucleus.

1916

Niels Bohr begins to develop the Copenhagen interpretation of quantum theory about protons and electrons.

1919

Arthur Eddington verifies Einstein's prediction by observing, during an eclipse, how rays of light from a distant star passing close to the Sun are bent by its gravity.

1924

Wolfgang Pauli discovers the Pauli exclusion principle which shows that no two electrons in an atom can be in the same quantum state.

1924

Prince Louis-Victor De Broglie discovers that all particles are essentially forms of wave.

1925

Max Born develops an influential new form of quantum mechanics.

1926

Paul Dirac predicts the existence of antiparticles and puts quantum theory on a firm basis.

1926

Erwin Schrödinger develops the theory of quantum wave mechanics, including Schrödinger's equation.

1927

Werner Heisenberg discovers the uncertainty principle which shows that you cannot determine exactly both the position and momentum of a particle simultaneously.

1932

Carl Anderson discovers the positron and the muon.

1932

James Chadwick discovers the neutron.

1946

Richard Feynman develops Feynman diagrams to help understand how electrons interact.

1947

Richard Feynman, Sin-Itiro Tomonaga and Julian Schwinger independently develop the idea of quantum electrodynamics (QED), to show how electrons interact with electromagnetic fields.

PAUL DIRAC

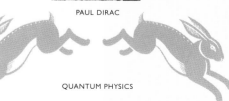

QUANTUM PHYSICS

1950

Freeman Dyson unifies the ideas of QED into a single theory.

1962

Murray Gell-Mann introduces the quark.

UNCERTAINTY

1967

Steven Weinberg and Abdus Salam show that there are four basic forces in physics: gravity, electromagnetism and the strong and weak nuclear forces.

1974

Sheldon Glashow creates the idea of quantum chromodynamics (QCD) to explain electromagnetism and the weak interaction between nuclear particles.

1982

John Schwarz and Michael Green develop the idea of cosmic strings.

ONE HOLE EXPERIMENT

Space and Time
ASTRONOMY

Since the human race gained the faculty to wonder, we have gazed at the beauty of the stars. Our curiosity has been drawn, moth-like, to these incandescent nuclear candles, which are so far away that their now-ancient light is rendered feeble by a small, lonely moon. Initial wonder inevitably led to the development of one of the first sciences, astronomy – a science that would ultimately provide explanations of the firmament, the wandering stars, where we came from and how we are going to end.

ABOVE **The modern era of astronomy began with the invention of the telescope in the early seventeenth century.**

LEFT **The Universe is dotted with countless billions of stars. Yet until the 1950s, our knowledge of the way they work was merely theory. Now computers provide a powerful tool for simulating their internal workings.**

STAR STRUCTURE

Of all the fields within astronomy, stellar astrophysics must have seemed the one least likely to succeed. It explores the make-up and evolution of stars – objects that are so faint and far away that they cannot be observed in detail by even the most powerful ground-based or space-borne telescopes. Fortunately we do have a rather ordinary star to hand – the Sun – from which we can learn a great deal about the physical processes that influence the structure and lifecycles of its more remote relatives. Yet the Sun itself keeps a close guard on its secrets. Everything we know about its internal make-up must be theory, for its intimidatingly hot atmosphere means that there is no hope of delving below its surface with a space-probe in the foreseeable future. So stellar astrophysics was a huge, almost impossible challenge until the development, in the 1950s, of computers capable of simulating the interiors of both the Sun and stars.

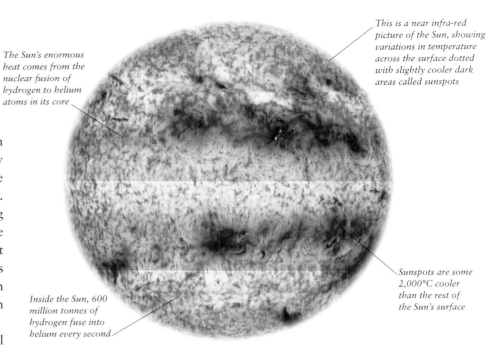

The Sun's enormous heat comes from the nuclear fusion of hydrogen to helium atoms in its core

This is a near infra-red picture of the Sun, showing variations in temperature across the surface dotted with slightly cooler dark areas called sunspots

Inside the Sun, 600 million tonnes of hydrogen fuse into helium every second

Sunspots are some 2,000°C cooler than the rest of the Sun's surface

SOLAR ENERGY

The first theories about the nature of the Sun were primitive by today's standards, but they addressed the first solar and stellar puzzle to face science: the source of the Sun's energy. Anaxagoras, an Ionian Greek philosopher living in about 400 BC, claimed that the Sun was made up of hot iron continually emitting heat and light towards Earth. Much later the early Victorians thought the Sun was powered by coal. Both theories seem ridiculous now, but the nuclear truth would probably have seemed ridiculous then.

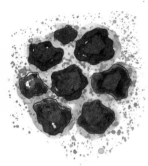

ABOVE **Astronomy has come an enormous way from the 19th century when it was believed that the heat of the Sun came from glowing coal.**

The Sun emits a colossal amount of radiation into space, and according to geological evidence it has been doing so since the formation of the Earth. The challenge was to find out where it came from. An early suggestion was that part of the Sun's gravitational, or more correctly potential, energy was converted into heat and then into radiation. Over a century ago Hermann von Helmholtz (1821–1894) and William Thomson, Lord Kelvin (1824–1907), put forward the idea that, owing to its enormous mass, a slowly contracting Sun would release enough potential energy to keep shining at its present luminosity for millions of years. This fitted with the contemporary belief that the Sun and the Earth were tens of millions of years old. However, subsequent geological work, using natural radioactivity to date rocks, led to a reassessment of the Earth's age. It was clearly hundreds of millions of years old at least, and perhaps billions. Gravitational energy could not power the Sun for so long, so what was the true energy source?

In 1905 Einstein put forward his special theory of relativity. This was the key that stellar astronomers were waiting for, for contained within the theory was the now famous equivalence relationship between mass and energy, $E = mc^2$.

ABOVE **The Sun at the centre of our Solar System is incredibly hot, with a temperature at the centre of over 15 million °C. A grain of sand heated to this temperature would set fire to everything within 100 kilometres of it.**

HERMANN HELMHOLTZ
1821–94

One of the most versatile scientists of the 19th century. Making major contributions in both science and the arts, Helmholtz suggested that the Sun's heat comes from the way its enormous mass makes it contract slowly, releasing energy all the time.

This suggested that mass might be being converted to energy, both in the heart of the Sun and in all stars. But how did the conversion come about?

In 1938 Hans Bethe of Cornell University suggested that the fusion of hydrogen nuclei into helium nuclei might be the energy source for stellar bodies. When four hydrogen nuclei are fused together to form a helium nucleus, the helium's mass is around 0.7 per cent less than the total of the four separate hydrogen nuclei. The missing 0.7 per cent is converted into energy, so that for every kilogram of hydrogen converted, 60,000 joules of energy are released. (A joule of energy is roughly the amount of energy acquired by an apple falling off a table by the time it hits the floor.) To satisfy the power output of the Sun, some 600 million tonnes of hydrogen must undergo fusion into 595 million tonnes of helium every second (releasing the energy equivalent of five million tonnes of matter in the process). Massive though this number may seem, it is dwarfed by the mass of hydrogen available in the Sun, which is enough for at least 10 billion years.

Although the principal source of the Sun's energy is the fusion of hydrogen and helium, we now know that several other thermonuclear reactions occur within the dense, high-temperature depths of stars. These produce carbon, oxygen and heavier elements. These are all important to the various stages of stellar evolution, and for the creation of planets in subsequent solar systems.

HUBBLE'S NEXT GENERATION

Arguably the most wondrous discoveries in astronomy in the twentieth century have been provided by the flagship mission of NASA's Great Observatories programme – the Hubble Space Telescope. It has, at least, returned the most fantastic pictures yet seen of everything from Comet Shoemaker-Levy 9 plunging into the atmosphere of Jupiter to the births of stars in stellar nebulae. Floating above the atmosphere, the telescope has extended humanity's vision beyond the visible portions of the electromagnetic spectrum and expanded our knowledge of how the Cosmos was born.

As the results from Hubble continue to roll in and inspire awe, astronomers are already planning the next generation and ways of picturing planets around other stars. If everything goes to plan the Next Generation Space Telescope will be launched in the year 2007 and operate through to 2017. The telescope will not only peer deep into space, but also further back in time, to when the galaxies were young. No time machines are involved in this trick, just the exploitation of the speed of light.

The more distant a light source in the Universe, the longer its light takes to reach us. So the light that reaches us today from such a source was emitted long ago, and the distance gives the difference in time. Light left a star one billion light years away one billion years ago. Light left the most distant objects visible over 13 billion years ago, way back in the dawn of the universe.

To help us further, the older and faster objects have their radiation 'redshifted' along the electromagnetic spectrum. An unfortunate side effect is that this also pushes them into an area of the spectrum that is absorbed by the atmosphere. That's a price worth paying though, because the redshift acts as an indicator of the speed of the object, and therefore of its age – a time stamp to tell us when the light was generated.

ABOVE **The Hubble Space Telescope has allowed us to see the very birthplace of stars in the Eagle Nebula. Revealed within the cloud of the nebula are globules called EGGs (evaporating gaseous globules) where stars begin life.**

The reason for this is related to the geometry of the Universe. When an object travels away from us – and everything in the Universe is still flying apart from the Big Bang – its light waves are stretched out and pushed along the spectrum towards and beyond visible red. This stretching is bigger for objects that are further apart. If you double the size of the Universe, two galaxies that are a million light years apart become two million light years apart, but two galaxies that are already two million light years apart become four million light years apart.

So the more distant objects in the Universe have bigger redshifts, regardless of where you measure from. This was discovered by Edwin Hubble at the end of the 1920s, and it was this discovery that showed the Universe is expanding.

Hubble's namesake space telescope could see a portion of the electromagnetic spectrum that gave it a window on the Universe from around four to nine billion years after the Big Bang. The Next Generation Space Telescope will have a wider spectral range near infra-red, and will be able to gaze back to a time when the Universe was young and the first stars and planets were forming at around 100 million years after the Big Bang. It will trace the evolution of entire galaxies through to the present day, by looking at many snapshots in the family albums of different galaxies. This is rather like trying to figure out how humans develop by looking at photographs of different people at different ages.

We know how galaxies look now, but precious little about how they came to be this way. If someone tells you they know how they formed, they are probably just quoting Edwin Hubble's theories on galactic evolution from the 1930s. A couple of generations on from his work, the Next Generation Space Telescope hopes to prove them.

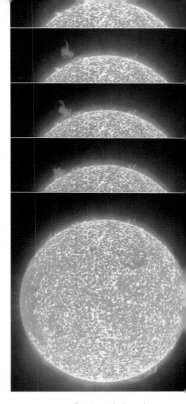

INSIDE THE SUN

The energy problem was solved, but on its own this was not enough to predict the future of the Sun. Nor did it tell us much about the interior of the Sun – something that we will never be able to find out by physical probing.

In the mid-1930s Sir Arthur Eddington, a British mathematician and astrophysicist, tried to develop theoretical methods to investigate the Sun and stars. A simplification of Eddington's work shows that a ball of gas can be used as a model of the Sun. It is in fact a plasma, the fourth state of matter where nuclei and electrons are separated into a conducting fluid of elementary particles, but Eddington relied on the fact that stellar plasma behaviour closely approximates to the behaviour of an ideal gas. The particles in a hot gas – or plasma – are in rapid motion and frequently collide with each other. These impacts are the source of gas pressure, which increases with particle temperature. This is combined in what is called the perfect gas law, which states that the pressure in a gas is proportional to the product of the density and temperature of a gas.

The tremendous gravitational forces within the plasma of a star should collapse the star towards its centre, but as stars remain broadly the same size for billions of years these forces must be balanced by another force pushing back out. Eddington assumed that this balancing force was the gas pressure of the plasma below. If this were the case the star would continue to collapse until the gas pressure was enough to hold up the weight of plasma above. If the gas pressure were greater than the gravitational forces, the star would expand and the pressure would drop.

One result of this is that the pressure, and therefore the density and temperature, of the star must increase with depth to support the increasing amount of matter pressing down from above. The condition of balance is called hydrostatic equilibrium, and by assuming that the Sun is in equilibrium Eddington was able to start calculating the internal temperatures and pressures of our star at various depths.

By the end of the 1930s we were able to calculate the internal conditions of the Sun, and other similar stars, with more confidence than we could calculate internal conditions of the Earth. The equations told us that pressures of over a billion atmospheres existed at the centre of the Sun, and temperatures of over ten million degrees centigrade. It is these tremendous temperatures and pressures, caused by the gravitational collapse of stars, that trigger thermonuclear fusion when a star initially collapses out of a stellar nebula.

STELLAR EVOLUTION

As the world of astronomy began to understand how the Sun currently worked, attention switched to how it evolved and how long it would last. Our fleeting perspective hinders our understanding of the ages of stars. Individually each of us lives for around a century at most, and our civilizations have been evolving for only a few thousand years. Yet the Sun has been pouring out energy for almost five billion years and will probably last for as long again. How could we hope to understand stellar evolution through epochs so beyond our own experience? The solution to this seemingly intractable quest – to match our observations of far-off alien objects in the Universe to our theories of the evolution of the Sun and stars – again came with the advent of computers capable of holding a model of stellar cores in their RAM chips.

ABOVE **Our knowledge of the Sun is certain to increase significantly in the near future as it is closely observed by a new generation of telescopes in space. This sequence is taken from the SOHO space telescope which was launched in 1996 and now hangs in space a million miles towards the Sun from the Earth.**

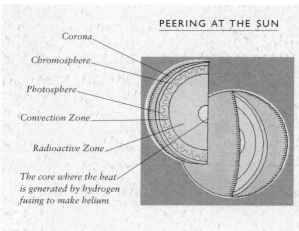

PEERING AT THE SUN

Corona
Chromosphere
Photosphere
Convection Zone
Radioactive Zone
The core where the heat is generated by hydrogen fusing to make helium

• The visible surface of the Sun (first yellow layer in diagram left) is called the photosphere.

• Above it is a bright layer of gas called the chromosphere.

• Beneath the surface are three main regions – the dense core, the radiation zone, where heat moves up towards the surface, and the convection zone where heat circulates.

Imagine a distant alien race snatching a few thousand human images from a burst of radio signals, imperfectly received through the distortion of the interstellar medium. The biologists among the aliens are curious: how do such unlikely creatures reproduce, and how do they evolve? From a few thousand images they sort them by scale and size. Do they waste away from large creatures to small pink ones? How do they change colour and features? Why do some have different body features – do they undergo some sort of metamorphosis? Science has acquired a number of similar snapshots of the stellar population of the skies, and has been on a similar quest to fit them into families and timelines.

RIGHT **One of the great space projects of the twenty-first century will be the construction of the International Space Station. This gigantic structure over 100m long will be put together piece by piece in space over five years.**

THE INTERNATIONAL SPACE STATION

As the twenty-first century begins, humankind will have a new permanently crewed science outpost orbiting the planet. Following on from space station Mir, the International Space Station will be the largest man-made structure in orbit. Together with its solar panels and trusses the ISS will be over 100 metres long and 70 metres wide.

The station will be slotted together from North American, Russian, Japanese and European parts in a 410-tonne international jigsaw puzzle, using pieces that have never been assembled on Earth. It will take 44 assembly flights and over five years to build, creating an orbital research complex that will carry a crew of seven scientists and engineers. They will study the near-Earth space environment and zero-gravity effects, with the hope of providing insights into how the human body works.

Without the oppressive force of gravity the body adapts in orbit. The first noticeable effect is that astronauts get 'chicken legs'. On Earth, the blood circulation system is a balance, between the drag of gravity and internal pressures of the skin and muscles. Homeostasis, the automatic internal maintenance of the body's environment, keeps that balance: sensors in the neck and brain detect pressure fluctuations and, in response, blood vessels contract and dilate or the kidneys extract urine from the system. In orbit this is all upset because gravity does not pull blood down to the legs, and the upward pressure of the blood vessels and tissues push too much upwards. As a result astronauts' legs become thinner – the chicken-leg effect – and their faces take on a puffy appearance. Experienced astronauts profess to 'flying dry' at launch. As pressure sensors in the neck detect an apparent excess of blood in the upper body, the first thing an astronaut wants to do in space is urinate (which, without gravity, is a technological challenge in itself).

Osteoporosis, the loss of calcium bone mass, affects many Earth-bound females after menopause and some unfortunates afflicted with bone-loss disease. Bones are normally constantly renewed as two groups of cells, osteoplasts and osteoblasts, move around the bones removing and laying calcium. For some reason the bone-laying cells slow down in space, and bones waste away. So astronauts exposed to zero-gravity for long periods suffer from osteoporosis. No one knows why for sure. The dream of space medical researchers is to use astronauts on the ISS to learn about zero-gravity bone loss, find a way to stop it, and use the same therapy to cure Earth-bound sufferers.

Materials can be manufactured in orbit that could never be made on Earth. Some materials – notably large protein molecules – collapse under their own weight when large numbers are grown. This frustrates medical researchers who would like to grow entire crystals of proteins – involving millions of molecules – so that the structures of the molecules can be investigated using X-ray crystallography. Zero-gravity conditions provide an ideal means of doing this, and as a consequence a few scientists orbiting the planet have the chance to achieve results that will be of lasting benefit to all humankind.

ABOVE **Most stars are huge for much of their lives. But once they have burned up all their nuclear fuel – or if they never really get going – they collapse to dwarf stars not much bigger than a planet.**

EJNAR HERTZSPRUNG
1873–1967
Hertzsprung, almost simultaneously with Henry Russell, spotted the connection between the absolute brightness of a star and its colour (which helps us work out its distance).

RED GIANTS

In 1905 the Danish astronomer Ejnar Hertzsprung compared stars by plotting their luminosity against their colour – which is in essence their surface temperature – on a graph or diagram. Coincidentally the same line of research was followed, quite independently, by American astronomer Henry Russell in 1913. The resulting Hertzsprung-Russell diagram – eventually completed in 1914 – has become the single most important tool in stellar astrophysics for deciphering the course of a star's life.

The most significant finding of Hertzsprung and Russell's work is that the stars are not randomly distributed over the diagram, as they would be if they exhibited all combinations of surface temperature and luminosity. Instead they cluster into certain parts of the diagram. The majority of stars follow a narrow diagonal line from top left to bottom right, called the main sequence, but a substantial number of stars cluster towards the top right towards the red end of the spectrum, showing they are both bright and cool.

This appeared to be a paradox. A low-energy, cool object normally emits much less light than a high-energy, hot object, so these stars are brighter than they should be. In order for them to be as bright as they are, they must be giant stars: some 10 to 100 times the size of the Sun. A few stars, such as Betelgeuse in Orion and Antares in Scorpius, lie even further towards the top right on the Hertzsprung-Russell diagram, so they must be much bigger than the ordinary giant stars and have become known as supergiants.

WHITE DWARFS

The red giants have their counterparts on the other flank of the main sequence, towards the bottom left of the Hertzsprung-Russell diagram. Observations of Sirius – a hot, blue 'main sequence' star – had revealed an orbiting companion that was both hot and dim. Following the logic used to deduce the nature of the giant stars, astronomers determined that it was a small yet powerful star which they called a white dwarf. As these are intrinsically faint stars, they must be relatively close to be visible. This explained why the first white dwarf to be discovered was seen orbiting Sirius, which is only eight light years away – a stone's throw in the Milky Way.

The orbit of the two stars about each other enabled the mass of the dwarf to be calculated, and its luminosity gave its radius and so its volume. The results were puzzling, because the figures gave the dwarf a density of over 100,000 times that of the Sun. Sir Arthur Eddington was one of the perplexed. As he noted,

'The message of the companion of Sirius when decoded ran, "I am composed of material 3000 times denser than anything you've ever come across. A tonne of my material would be a little nugget you could put in a matchbox." What reply could one make to something like that? Well, the reply most of us made in 1914 was, "Shut up: don't talk nonsense."'

More white dwarfs were discovered, however, and there are now known to be hundreds of them. But how did they develop? Cool 'protostars' are

DOUBLE STARS

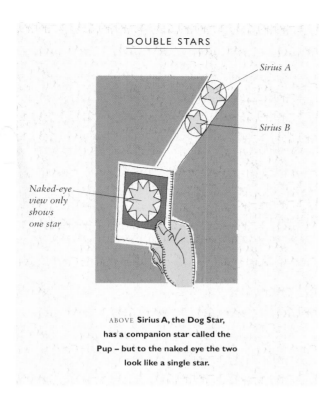

Sirius A

Sirius B

Naked-eye view only shows one star

ABOVE **Sirius A, the Dog Star, has a companion star called the Pup – but to the naked eye the two look like a single star.**

known to form from the collapse of gas nebulae into regions where, eventually, the pressure and heat ignite hydrogen fusion. As this happens they enter the main sequence on the Hertzsprung-Russell diagram. Theoretical calculations and simulations have completed the sequence of events from that point onwards.

Once on the main sequence a star is almost totally powered by the conversion of hydrogen to helium. It stays there for most of its life, but eventually – after only ten per cent of the star's mass has been converted – the huge stocks of hydrogen in the star's core become depleted. The core is now starved of its energy source, even though hydrogen fusion continues farther out in the star, and without any outflow of energy to support its mass the core contracts.

As the star's core shrinks it releases gravitational potential energy. This is absorbed by the outer layers of its atmosphere making them expand to huge proportions. The star now produces more energy than before, but because it has a bigger surface the amount of energy for the surface area is less. So the surface becomes cooler and redder as seen from the outside. The star has moved up and to the right of the Hertzsprung-Russell diagram, into the region of the red giants.

WHAT HAPPENS NEXT?

The answer came from astrophysicists of the 1950s puzzling over why we are here at all. In particular, astrophysicists such as Sir Fred Hoyle wanted to know the source of the elements from which we are made. All the elements were made by the fusing together of atoms of lighter elements as the universe evolved. But Hoyle wanted to know just how this happened.

It was known from particle physics that the elements you might hope to make by sticking pairs of helium nuclei together were unstable, so they could not be made inside stars. But if you could somehow make carbon, it would be relatively easy to make even heavier elements by adding helium nuclei (also known as alpha particles) to them.

Hoyle realized that this could occur only if an unusual process, called a resonance, happened when three helium nuclei collided, in what became known as the triple-alpha process. He persuaded nuclear physicist William Fowler to carry out a search for this kind of resonance using the particle accelerator at Caltech. It was duly found – Hoyle had used his understanding of stars to predict how particles behave at high energy. Together with Geoffrey and Margaret Burbidge, Hoyle and Fowler built from this result a complete theory of how the elements are manufactured out of hydrogen and helium inside stars. Their calculations had a built-in confirmation too, for the theoretical ratios of the elements produced are just about spot-on when compared with nature.

The process has terminal ramifications for the red giants. The triple-alpha process begins abruptly in most stars, taking place throughout their cores in a rapid release of energy. The energy reverses the shrinkage of the stellar core and the outer layers of the star collapse inward, heating up as they go – ending the first red giant phase after around a billion years of bloated existence. Then, over a few hundred million years or less, the star expands to red giant stage again and again as new nuclear processes fuse higher elements and then exhaust their fuel – shifting the star to left and right across the Hertzsprung-Russell diagram.

ABOVE **This is an optical photograph of one of the most beautiful objects in the night sky, the Pleiades open star cluster (M45) in the constellation of Taurus. Visible are the clouds of interstellar gas and dust – remnants of the original nebula from which the stars formed.**

SUPERNOVAS AND BLACK HOLES

Sometimes, if the star is a massive supergiant and its core has been almost totally converted to iron, it can go supernova. The core suddenly collapses, releasing huge amounts of gravitation energy, and the outer layers of the disrupted star are sent hurtling out into space. These explosions eject their matter into the interstellar medium, enriching the material that will make the next generation of stars with the elements that form rocky worlds – and, on at least one such world, our own planet Earth, life.

In 1987 the light from a star going supernova reached us from a neighbouring galaxy over 1,175,000 light years away. It was a brilliant confirmation of theory. Astronomers had predicted that the collapse of a core would produce a massive number of neutrinos – massless particles which interact only weakly with matter. Just before the event it was all still theory, but at 07:35:42 GMT on 27 February underground detectors registered increased numbers of neutrinos passing through the Earth. The following night a new brilliant object was seen – the death throes of a dying star.

These convulsions end with a star such as the 1987 supernova totally exhausting its store of nuclear energy. The dying star then collapses until it becomes extremely dense. What happens next depends on its mass.

The remains often form a white dwarf, in which the internal pressure of the star supports its outer layers. The light that a white dwarf emits is just the dying embers of its former self – the stored

RIGHT **Black holes are points in space so dense that their gravity sucks everything in, including light. Once they were a purely theoretical construct – the logical end-point of the collapse of a dying star under its own gravity. But there is increasing evidence that such objects exist at the heart of many galaxies – including our own Milky Way.**

THERE WAS A SILENCE, A BANG AND THEN MAYBE A BIG CRUNCH

In 1912 Vesto Slipher made a discovery that laid the foundation for our modern understanding of cosmology. Under the direction of Percival Lowell he obtained the first spectral analysis of the light coming from an astronomical object which looked like a spiralling cloud of gas: a nebula. Lowell believed the nebula to be a solar system in the process of forming – the distance and scale of the cloud being then quite unknown.

But the spectrum of the nebula turned out to be nothing like Slipher expected from a young solar system. To his surprise the spectral lines of the elements in the light showed a large Doppler shift, indicating that it was receding at about 300km per second. Slipher continued the work on other, similar objects over the next two decades, and found that all but a few were moving at speeds of at least 2,000km per second, and most were going away from us. These speeds showed that the clouds were not in fact members of our galaxy – they were other galaxies, spirals not of gas and dust particles, but stars.

Meanwhile Henrietta Leavitt of Harvard College Observatory had been analyzing the changing intensities of light, or light curves, emitted by stars called Cepheid variables. The stars she was studying were all known to lie at roughly the same distance from Earth because they were in the same large galaxy. As Leavitt plotted the periods of the variation in their light curves against the amount of light they emitted, the relationship became obvious: the longer the period of the rise and fall of the light curve, the brighter the Cepheid variable.

Stars generally appear dimmer the further they are from Earth, because the light is spread over a larger and larger sphere the further it gets from the star. The sphere size is in fact proportional in size to the distance squared. If you know the original light output, though, you can work backwards from the observed brightness and figure out the distances involved. The periods of the Cepheid pulsations gave that original brightness away, enabling these stars to be used as cosmic distance markers.

Edwin Hubble, working at the Mount Wilson Observatory in 1923, recognized that light curves in the analysis of light from some of Slipher's nebulae showed evidence of Cepheid variables, so he used these yardsticks to assess the distance of the nebulae. The Cepheid variables are rare supergiant stars, and yet the ones in the nebulae were extremely faint. Their periods suggested they should be much brighter, so the objects had to be incredibly remote. The 'nebulae' had been established as galaxy-sized objects lying far beyond our own, in deep space.

Slipher's work on the redshifts of galaxies, and their speeds as they move away from us, was built upon by Hubble with more spectral observations and determinations of distance using Cepheid stars. It soon became apparent that the further away an object was, the more redshifted it was and the faster it receded

heat that, after billions of years, is exhausted so the now planet-sized body with the mass of an entire Sun fades to a cold body called a black dwarf.

White dwarfs are formed from matter that is incredibly compressed, so the normal rules for matter no longer apply. The interior is a degenerate plasma in which the atoms have been separated into electrons and ions; they are squashed into a much smaller volume than normal until electrostatic forces support the mass above. The more mass on top, the more the material is compressed, with the result that the final dimensions of the white dwarf run contrary to common sense: the greater the mass of the original star, the smaller the radius of the resulting white dwarf is going to be.

from our position: everything in the Universe was flying apart. The result was published in a classic paper in 1931 and it became known as Hubble's law.

Hubble's law enables us to work out the age of the Universe and speculate about its ultimate fate. Hubble's law is simply expressed in mathematics as V = HD, with V being the radial velocity of a galaxy and D the distance to it. H is the Hubble constant derived from observation and indicates how fast a galaxy is travelling per unit distance away. Assuming the Universe started in the Big Bang – the point when all matter and spacetime exploded into existence – and the galaxies have all been travelling away at a constant speed since then, the velocity (V) of any object in the Universe is the distance it has travelled (D) divided by the age of the Universe. Combining this with Hubble's equation and cancelling the distances, the result is that the age of the Universe is just the reciprocal of the Hubble constant, putting the Universe at 15 billion years old.

GRAVITY VERSUS SPEED

If a weight is given a certain push on the Moon, it will fly away from the surface continuously

On a larger body like the Earth, the same push could put the object into orbit

But on a still bigger planet, like Jupiter, exactly the same push would see the weight rise a little way before dropping to the surface again

And when will it all end? The explosive expansion of the Universe is slowing down as its own gravity steals kinetic energy from its matter. How much it slows down relies on how much gravity there is and the total mass of the Universe. If the density of matter in the Universe is below 11 hydrogen atoms per cubic centimetre, the Universe will expand forever. Everything will slowly lose energy in radiation and the galaxies will drift apart into infinity. If the density of the Universe is exactly this number, the Universe will expand up to a point where things have slowed until they stand still, becoming a static collection of distant galaxies.

If the Universe is denser than 11 hydrogen atoms per cubic centimetre, even if it is only by just a few atoms, the Universe will eventually stop expanding, and begin to return towards the point of the Big Bang. At this point, all of the Universe will occupy the same region of space in what the Princeton physicist John Wheeler has dubbed 'the big crunch'.

A GHOSTLY WIND AND LIGHTS IN THE SKY

Awesome and beautiful though the stars and planets may be, they are arguably upstaged by far rarer lights in the sky – the dancing curtains of pulsating blue and green light known as the aurora borealis, or northern lights. Science struggled to explain this phenomenon for centuries. Galileo proposed that the aurora was caused by air rising out of Earth's shadow and being illuminated by the Sun, and Descartes believed the lights to be reflections from high-altitude ice crystals.

Scientists began to feel their way towards a correct explanation of the northern lights in 1722 when George Graham, a famous London instrument maker, noticed that his incredibly accurate compasses were continually being disturbed by little perturbations. A Swedish team confirmed his observations in 1740, and noticed that the magnetic disturbances were most violent during large auroral displays. So the lights in the sky were linked to the Earth's magnetic field.

Why the field was so disturbed and how it lit the sky were both unknown until early nineteenth-century observers began to measure the disturbances while counting the dark patches on the surface of the Sun, known as sunspots. The intensity of the perturbations varied with the sunspot numbers, so somehow the Sun was disturbing the Earth's magnetic field and lighting the sky at night.

The evidence implicating the Sun became irrefutable following a chance observation by an English astronomer called Richard Carrington. On 1 September 1859 Carrington was sketching a number of sunspots when a transient incandescent flare of light from the Sun startled him. By the time he fetched a witness the event had all but faded away, but fortunately another observer had seen it. Furthermore Kew Observatory in London had made simultaneous measurements showing that the magnetic field had been affected almost instantaneously. Finally, within 18 hours, one of the strongest magnetic storms ever recorded broke out, with auroras stretching as far south as Puerto Rico. Whatever was coming from the Sun to cause the aurora was travelling at nearly 2,300km per second.

It was not until 1918 that the solar-terrestrial link was solved. Sydney Chapman postulated that a negatively charged beam of electrons might be streaming out of the Sun in a 'solar wind' towards the Earth to disrupt its magnetic field and penetrate its atmosphere, exciting the molecules of the air to fluoresce in the greens and blues of the aurora. Chapman's idea did not survive in its original form, for surely, said the critics, such a beam would destroy itself through mutual electrostatic repulsion of the negatively charged particles? But a modification proved a breakthrough – why just electrons, why not a plasma? A plasma is a hot soup of negatively charged electrons and positively charged ions, with equal numbers of each so that the plasma itself is electrically neutral and not auto-destructive.

The problem was solved, and the plasma idea carried an implication that was proven by spacecraft years later. Plasmas carry magnetic fields as they move, and so the solar wind plasma could shape the Earth's magnetic field. As the solar wind streams by it compresses the front of the field and stretches it out behind the Earth to create a tail, forming a magnetic cavity around the Earth that the solar wind cannot penetrate.

Soon observations of comets with streaming tails were providing clues as to the density of the solar wind and its speed. As comets orbit the Sun they leave tails of gas and ice, but the solar wind deflects the tails so that they point away from the Sun. To deflect the comet tails so much the wind had to be moving phenomenally fast, at an average of around 450km per second, and be incredibly rarefied, with less than 30 particles per cubic centimetre. If you were to cup your hands in the wind on Earth you would be holding over 300 million trillion particles. Since the solar wind is so rarified, if you were to stand in it – even at the windspeed of 450km per second – it would not have the momentum to move a single hair on your head.

Rockets soon confirmed this. In 1958, a year after Sputnik 1, the US space programme was inaugurated with the launch of the Explorer 1 spacecraft. It carried a geiger counter supplied by Professor James Van Allen, a physicist from the University of Iowa. The instrument discovered that the Earth's magnetic field not only protects us from the worst of the solar onslaught of radiation, but also marshals some particles in lethal belts of radiation about the planet.

ABOVE **The spectacular glowing lights of the aurora borealis are created by the interaction of the charged particles of the solar wind with Earth's atmosphere.**

BLACK HOLES

In 1930 Subrahmanyan Chandrasekhar, who pioneered theoretical studies of white dwarfs, realized that there must be an upper limit to the forces that electrons could provide in attempting to support increasing loads. Chandrasekhar's limit showed that, for a star exceeding 1.4 times the mass of the Sun, the interior of the resulting white dwarf would be incapable of supporting the mass of its outer layers. These would therefore continue to collapse, so the white dwarf would be close to having no size whatsoever. This of course defies common sense – and strays into a realm of objects where the usual laws of physics begin to struggle. The answers have yet to be fully worked out and confirmed, but as more massive stars collapse, it is suspected that they become black holes.

Armed with only the conventional Newtonian theory of light, earlier astronomers such as John Michell and Pierre Simon, Marquis de Laplace, had proposed the existence of such 'dark stars'. In 1796 Laplace speculated that massive stars might exist with so great a gravitational attraction that light would bend around on itself and never escape. In 1967 John Wheeler, a Princeton University physicist, considered Laplace's 'dark bodies' in the context of Einstein's general relativity theory and dubbed the objects black holes.

All bodies such as stars have an escape velocity – the speed that an object must achieve to overcome the body's gravity and leave it permanently. A rocket launched from Earth, for example, must achieve Earth's escape velocity before it can head out into space. The escape velocity is related to the mass of the body and its radius. As the mass goes up – or as in the case of a collapsing star the radius shrinks – the escape velocity increases. In a black hole, the radius of the star has shrunk below the point where the escape velocity is equal to the speed of light, so all light from the star is permanently trapped.

Nothing quite as exotic will happen to our Sun. Safely under the three solar mass minimum requirement for a black hole, the Sun will probably end its days as a white dwarf after swelling to a red giant. A good guess is that the Sun has another five billion years or so to go before its hydrogen fuel becomes depleted and a pure helium core forms. At that stage it will evolve into a red giant with its outer layer engulfing everything as far out as the orbit of Venus – and incinerating the pale blue dot of Earth in a red inferno. The stars that created the matter for life, and provide the energy to power it, will finally eradicate all traces of it.

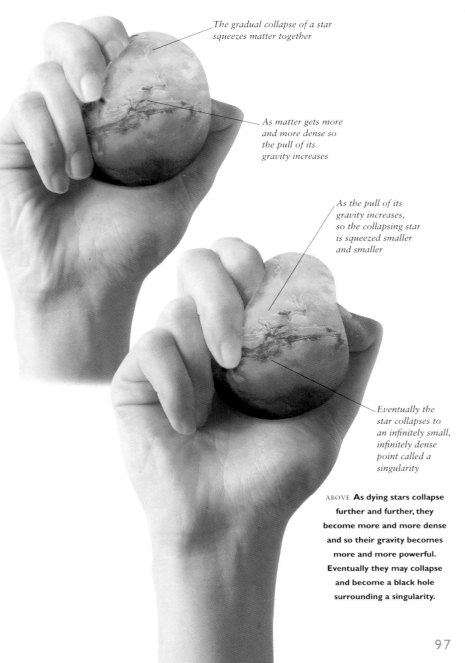

BELOW **When stars begin to burn out, there is no heat to push them outwards and so they begin to collapse under the power of their own gravity.**

The gradual collapse of a star squeezes matter together

As matter gets more and more dense so the pull of its gravity increases

As the pull of its gravity increases, so the collapsing star is squeezed smaller and smaller

Eventually the star collapses to an infinitely small, infinitely dense point called a singularity

ABOVE **As dying stars collapse further and further, they become more and more dense and so their gravity becomes more and more powerful. Eventually they may collapse and become a black hole surrounding a singularity.**

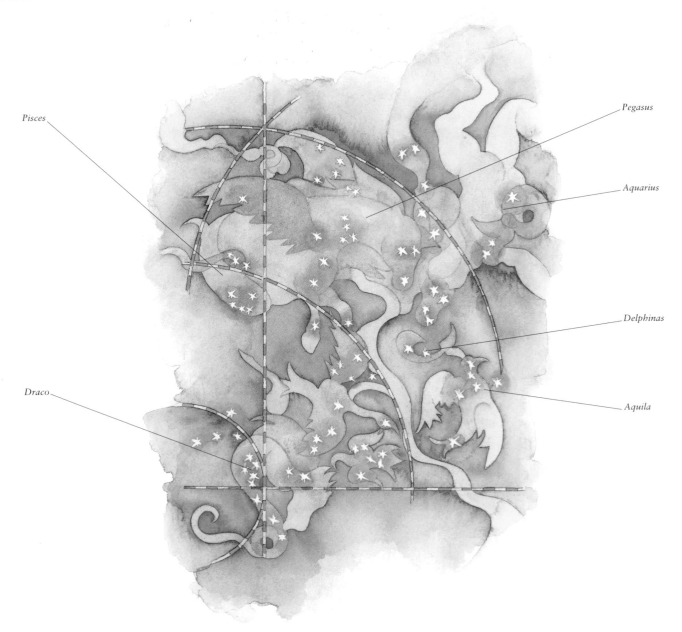

Pisces

Pegasus

Aquarius

Delphinas

Draco

Aquila

THE PLANETS

The development of planetary astronomy has been an evolutionary process. Theories and ideas to explain the nature of the Moon and the wandering lights in the sky have been mutating in our minds ever since history began. The theories born of scientific conjecture have enjoyed lives of varying length, and not always in accordance with their validity, since the natural selection of argument and debate has sometimes been impeded by the scientific politics of the time. And occasionally a random scientific cataclysm, such as the invention of the telescope or the advent of spacecraft, has wiped out the dominant theories and cleared the field for new breeds of ideas.

ABOVE **Since ancient times, astronomers have found their way round the night sky by grouping the stars into familiar patterns called constellations, each with its own name, typically based on a Greek myth. There is actually no connection between the stars in a constellation; their proximity in space is purely visual and coincidental.**

THE EGOCENTRIC UNIVERSE

Planetary debate is littered with the remains of old scientific theories. The idea that the Earth was at the centre of the Universe, with the Sun, planets and stars spinning around us, is one of the more spectacular relics. In retrospect it seems unlikely, but it dominated thinking for most of history. It appealed to an egotism that placed us at the summit of creation. It also fitted in with everyday observation, provided one accepted a few mysterious phenomena. Such mysteries were routine in an age dominated by religious faith, and it was the refusal to accept mystery that enabled astronomers such as Copernicus and Galileo to lay the foundations of modern planetary science.

A BRIEF HISTORY OF THE SOLAR SYSTEM

A massive, cold, dark cloud of interstellar gas once lay where our Solar System is today. Hydrogen and helium formed most of this solar nebula – the matter from which the Sun and planets formed. Most of the rest of the matter, the heavier elements and molecules, was formed into tiny specks or snowflakes of ice. The solar nebula, some two or three times the mass of the Sun, existed in this form for an indeterminable age.

According to one theory this state was disturbed around 4.6 billion years ago by the massive explosion of a nearby star, reaching the end of its life as it ran out of fuel. The shock waves that pulsed through the gas cloud disturbed it into collapsing under its own gravity – the end of one star triggering the formation of our own.

This concept of the Solar System's origin appears to have been first suggested by the German philosopher Immanuel Kant, and was picked up by the French astronomer Pierre Simon, Marquis de Laplace in 1796. Laplace developed it mathematically into our current understanding of how the cloud collapsed. He postulated that the solar nebula must have been rotating slightly, and as matter collapsed inward its rotational speed increased – just as a pirouetting ice-skater spins faster as she pulls her arms in. Centrifugal forces prevented the majority of the matter that spun in the plane around the protosun from falling into the developing star, but most of the matter in the cloud above and below the plane collapsed inward. The process converted the shapeless cloud into a rapidly spinning disc – explaining why the planets all lie in the same plane.

Over nearly 100 million years the solar nebula continued to collapse, converting its gravitational potential energy into thermal energy. In time the centre of the cloud was heated and compressed to form a protosun. Temperatures within this newborn Sun quickly rose to 2,000 kelvin, while in the cold outer reaches the planetary nebula rarely saw the dizzying heights of 50 kelvin (around –223°C). This difference in temperature led to the formation of the small, rocky inner planets and giant, gaseous outer planets. As the inner Solar System heated, the icy snowflakes that had existed for billions of years evaporated, leaving only the less abundant rocky materials that built the terrestrial planets and asteroid belt. Meanwhile the ice crystals survived in the cooler outer portion of the contracting nebula, to coalesce into the giant planets which are now rich in such materials as ammonia and methane.

RIGHT **The American space shuttle brought about a new era in space exploration – making repeat visits to orbiting space stations and satellites a matter of routine.**

In the inner Solar System accretion formed the four terrestrial planets and several dozen Moon-sized objects. The dust grains in the planetary nebula bumped into each other over millions of years and were accreted by electric and gravitational forces into pebbles and rocks, and then eventually a billion or so planetesimals each around 10km in diameter. The planetesimals were large enough to attract their neighbours and began to smash into one another, welding themselves into proto-planets. The Solar System was quite wasteful during this period, for each time a growing proto-planet attracted a planetesimal it had a fifty-fifty chance of either absorbing it or gravitationally slinging it out of the Solar System.

NICOLAUS COPERNICUS

1473–1543

Copernicus was the first astronomer to realize that the Earth circles the Sun along with all the other planets.

THE PTOLEMAIC PANTHEON

For the early astronomers, only the most powerful of the gods occupied the sky. Save for the Sun, the fixed stars and the Moon, the Earth's only other companions were the wandering deities Mercury, Venus, Mars, Saturn and king Jupiter. An Alexandrian called Ptolemy, one of the great astronomers of antiquity, had fixed Earth at the centre of this planetary pantheon in about AD 140. His reasoning was straightforward. The Moon obviously circled the Earth and the Sun appeared to, daily, so why should the wandering stars be any different? Small sub-orbits of the planets, called epicycles, could be employed to explain away the strange periodic looping courses the planets sometimes took.

Ptolemy's system was eventually consigned to the fossil record of scientific theory, and the wandering stars – including Earth – are now known to circle the Sun. What's more, this Sun-centred system is now known to contain over 70 moons and planets, 3,450 detectable asteroids and countless interloping comets which form a swarming halo to the system. It's actually quite crowded up there, and we've even run out of small gods to name the planets after.

Asteroids are rocky lumps left over from the formation of the Solar System

A million asteroids are more than one kilometre across

The progress from a handful of gods orbiting the Earth to a crowded Solar System of varied fantastic worlds occurred in fits and starts, as did the technological progress that made it possible. The Ptolemaic pantheon of just five planets lasted until the end of the Renaissance, when our concept of the heavens first exploded.

A SPINNING EARTH

The explosive charge was laid by Nicolaus Copernicus, a Polish cleric and scientist. Born in 1473, he eschewed his legal and medical training to concentrate on his main interests of astronomy and mathematics. By the time he reached middle age he was a well-known authority on astronomy. He built on the Ptolemaic view of the Universe and quietly championed the idea that it was spherical, and that the motions of all heavenly bodies must be composed of uniform circular motions.

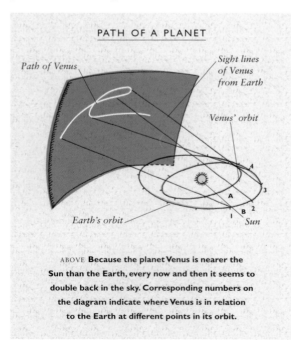

PATH OF A PLANET

Path of Venus

Sight lines of Venus from Earth

Venus' orbit

Earth's orbit

Sun

ABOVE **Because the planet Venus is nearer the Sun than the Earth, every now and then it seems to double back in the sky. Corresponding numbers on the diagram indicate where Venus is in relation to the Earth at different points in its orbit.**

Even a small asteroid crashing into the Earth (a meteorite) can do considerable damage

ABOVE **There are thousands of large lumps of rock in the Solar System circling the Sun called asteroids. Some are as large as a person, and others are much, much larger – the biggest, Ceres, is 1,000km across.**

However, crucially, he argued that the apparent annual motion of the Sun around the Earth could be equally well represented as the Earth moving around the Sun. He reasoned that the daily rotation of the celestial sphere of stars about the Earth could be accounted for by assuming that the Earth rotates about a fixed axis, while the celestial sphere is stationary. His critics were alarmed by this – would the Earth spinning once a day not rip itself apart? Copernicus' response was elegant. He answered that although such a motion might break the planet apart, the even faster motion needed to rotate the massive celestial sphere once a day – as required by the alternative hypothesis – would cause much more devastation.

Copernicus published his ideas in 1543, the year of his death, but at that point the theory was just one among many. There was no evidence to either prove or disprove it, so its explosive potential remained unrealized for over half a century.

SHOOTING STARS, ASTEROIDS AND COMETS

The Solar System is not quite completed, for the process of planet building has never ended. There is still a lot of raw material strewn about the place making it rather dangerous sometimes, but also filling the sky with wondrous displays.

There are now known to be about 3,450 detectable asteroids littering our cosmic back yard. Many are herded between Mars and Jupiter, but quite a few are drifting about the system. Once every 100,000 years one large enough to kill off an entire branch of evolution crashes into the Earth's atmosphere. Every 100 years one large enough to destroy a city explodes in the atmosphere or craters the planet's surface. So far we have been lucky – no city has ever been destroyed by an asteroid impact – but it could easily happen.

The comets are potentially deadly too. Forming a swarming distant halo around the Sun, they occasionally graze it closely enough to heat up and lose tonnes of volatiles in a tail that streams out behind – giving them a majestic appearance for many months that far exceeds their normal status in the Solar System. Sometimes they approach the Sun a little too closely, and swing around the far side never to return. Sometimes they run out of the gases and ices that fuel their displays and become little more than asteroids with eccentric orbits. Their careers in the limelight of the inner Solar System are short. Many wait billions of years in the distant Oort Cloud of comets surrounding the Sun before streaking across the skies.

Some comets end their lives crashing into the Sun or getting ripped apart by the huge tidal forces of Jupiter – as happened to Comet Shoemaker-Levy 9 in 1994. But some are more hardy stars of the sky, keeping a wary distance from both the Sun and Jupiter. Comet Halley puts on a variable performance every 76 years, and Comet Encke puts on its dull show every 3.3 years. By the time they cross the orbit of the Earth they are always shedding gas and dust. Earth crosses such belts of dusty matter every year, and as the dust dives into the upper atmosphere it burns up in meteor displays that, shortly after a comet's visit, can fill the sky with shooting stars.

FIERY MESSENGERS

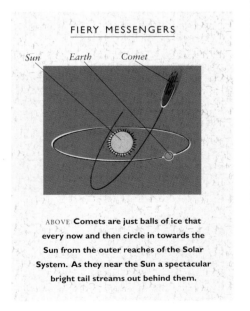

Sun   *Earth*   *Comet*

ABOVE **Comets are just balls of ice that every now and then circle in towards the Sun from the outer reaches of the Solar System. As they near the Sun a spectacular bright tail streams out behind them.**

HANS LIPPERSHEY

1570–1619

Lippershey was the Dutch lens maker who is credited with inventing the telescope in 1608 – so legend has it, after children playing in his shop tried looking through two lenses together.

ABOVE RIGHT **Like the balls on a juggler's stick, the planets circle round the Sun in the same orbit, held forever in balance between the pull of the Sun's gravity and their own velocity which tries to make them fly off into the darkness of space.**

THE FUSE IGNITED

In 1608 Hans Lippershey, a Dutch lens grinder and spectacle maker, created the refracting telescope. The Dutch government tried to suppress news of the invention – such was its military potential – but fortunately without success. The news leaked to Galileo Galilei, an Italian astronomer. Within a year, and without having ever seen an assembled telescope, he had developed it into a tool to change the way the world thought.

Holding his magnifying glass to the heavens, Galileo provided the evidence to eject the Earth from its position at the centre of the Universe. In 1610 he published what he had seen in a small book, Sidereal Messenger, and astonished the world. He detailed stars that had previously been too faint to see, and revealed that the nebulous blurs of the Milky Way and similar features were simply dense concentrations of distant stars.

Most importantly, perhaps, he found four moons revolving around Jupiter. This last discovery removed a plank that Ptolemy's defenders had relied upon to undermine Copernicus' ideas – for they had argued that if the Earth was in motion, the Moon would be left behind as it could not keep up with such a rapidly-moving planet. Galileo had shown that a centre of motion, Jupiter, could in turn be in motion and not lose the moons orbiting it.

The main confirmation of Copernican theory, though, was Galileo's observation of the phases of Venus. Armed with the enhanced observing power of his telescope, Galileo saw that Venus went through phases like those of the Moon, from new to crescent and through to fully lit. In order to explain this, Venus needed to be at times behind the Sun and at others between the Earth and the Sun. Since these facts could not be reconciled with the model of Venus orbiting the Earth, they finally extinguished Ptolemy's theory. As Galileo wrote in a letter dated 1615, 'in my studies of astronomy and philosophy I hold this opinion about the Universe, that the Sun remains fixed in the centre of the circle of heavenly bodies, without changing its place; and the Earth, turning upon itself, moves around the Sun.'

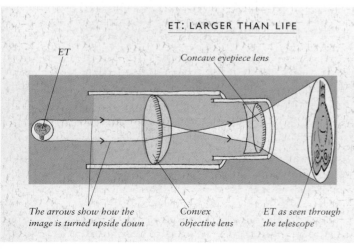

ET: LARGER THAN LIFE

ET

Concave eyepiece lens

The arrows show how the image is turned upside down

Convex objective lens

ET as seen through the telescope

LEFT **Like all astronomical telescopes the early telescopes used by Galileo showed an upside-down image. Ordinary telescopes for use on Earth have an extra lens for turning the image the right way round again.**

In 1616 Galileo was warned to change his views by the Church, which maintained that the Ptolemaic system was correct. But Galileo pressed on, and eventually his comprehensive demolition of the old theory convinced the Pope that Galileo was ridiculing him. He placed the astronomer at the mercy of the Inquisition, accused of heresy. He was tried in 1632, when he was 69, and forced into a public recantation.

Galileo spent the rest of his days under house arrest, but despite the Church's desperate rearguard action one of the greatest of all scientific battles had been won. By the end of the century Copernicus' Sun-centred or 'heliocentric' theory dominated scientific thinking, and the Sun had twice the complement of children it started the century with. Galileo had notched up four moons of Jupiter (Callisto, Europa, Ganymede and Io), Christiaan Huygens had discovered Titan, the biggest moon of Saturn, in 1655, and within 30 years Cassini had found four more moons orbiting Saturn.

BELOW **Titan, the largest moon of Saturn, was one of the many discoveries made once astronomers began to look at the night sky through telescopes. It was spotted by Christiaan Huygens in 1655.**

SPECTRAL WINDOWS ON OTHER WORLDS

To really get a good look at space you really need to go up there. Not just because you will be closer – that hardly matters on the scale of things – but because you will have removed around ten tonnes of eddying gas immediately above you that selectively filters and distorts light from the Universe. It was fortunate for the evolution of life that the atmosphere blocked most of the harmful cosmic emissions in the electromagnetic spectrum, but this does not help the quest of the astronomer. A treasure-trove of information stops dead at varying altitudes above us, never to reach the telescopes on the ground.

Each part of the spectrum reveals a new dimension of the Universe. Turning your spectral sights on the planets and their moons would betray the chemical composition of their atmospheres and surfaces, for the colour spectra of different molecules each have their own characteristic signature. If you were looking at the stars and galaxies in the infrared, the world of cosmology would open up to you with fainter, more distant, older galaxies shifted towards red as they recede from us.

One of the first space missions to see the shielded emissions of the Universe was a rocket flight in 1962, planned by a team led by Riccardo Giacconi, an American astronomer. The flight purported to be looking for X-ray emissions from the Moon, a potential radiation hazard for astronauts. It found none, but did see a new and extremely bright (in X-ray terms) object they named Scorpius X-1.

More rocket missions and orbital X-ray space telescopes quickly followed, allowing astronomers to see high-energy X-ray objects such as quasars and binary stars annihilating each other in the massive outpourings of energy required to generate X-rays. After the long years of atmospheric prohibition, dozens of X-ray missions are now showing observation in this frequency to be one of the most important fields of modern astronomy.

The SOHO solar space telescope launched in the early 1990s is held in place by a gravitational anomaly where the gravity of the Sun and Earth almost cancel, enabling it to sit permanently between the two bodies. From this unique vantage point the telescope gazes at the Sun, registering wavelengths that would never reach the ground. These show the turbulence of the solar atmosphere as it churns and explodes, jettisoning millions of tonnes of material into the Solar System.

The Hubble Space Telescope is a relatively new addition, launched in 1990. Above the distorting atmosphere it can see visual wavelengths from the stars with greater clarity than ever before, peer back billions of years on infrared wavelengths to see the oldest objects in the Universe, and receive ultraviolet emissions from some of the hottest bodies in the Cosmos.

So the dawning of the space age has allowed astronomers to rise above the atmosphere and, using 'eyes' that can see well beyond the limited range of visible light, gain new spectral windows on the Universe.

OBSERVATION AND CALCULATION

Once the easier targets had been picked off by the telescope there was a lull in the rate of discovery until the late eighteenth century, when William Herschel added five heavenly bodies to the total – including a new planet. In March 1781 the German-English musician and astronomer made a routine survey of the sky. Training his telescope on the constellation of Gemini, he noticed a star which did not have the pinprick appearance of surrounding stars and seemed to be a small disc of light. He followed its motion for weeks, believing it to be a comet, and plotted a solution for its apparent motion – that of an orbit lying beyond Saturn. Herschel had discovered Uranus.

Neptune, which lies beyond Uranus, was discovered in 1846 without even being seen. As the orbit of Uranus was observed and refined to better accuracy an odd fact emerged – its orbit did not fit with Newton's theory of gravity. Even allowing for the perturbing gravitational effects of Jupiter and Saturn, it was found that Uranus did not move on an orbit that could be squared with earlier observations. Two mathematicians, John Couch Adams and Urbain Jean Joseph Leverrier,

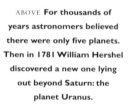

ABOVE **For thousands of years astronomers believed there were only five planets. Then in 1781 William Hershel discovered a new one lying out beyond Saturn: the planet Uranus.**

independently took this problem and speculated that an eighth, hidden planet might be pulling Uranus' orbit out of shape.

In 1843 Adams, a young Englishman, began an analysis of the perturbations of Uranus. Within two years he delivered a predicted position for the new planet to the Astronomer Royal. He was correct to within two degrees: a thumb's width when held to the sky. Unaware of Adams' work, French mathematician Leverrier did the same the following year. He sent his results to Johann Galle of the Berlin Observatory, and within a few hours of receiving the letter Galle compared old charts of the Aquarius region, where the planet was projected to be, with the night sky. He found the planet to be within a degree of the position predicted by Leverrier.

It is possible that Galileo saw the planet before this. Astronomical historians have calculated that Neptune lay in the background of the line of sight to Jupiter in 1613, a few years after Galileo had discovered the Galilean moons of the planet. Galileo's notebooks show that he had indeed seen a moving star – but did not realize the significance of what he saw.

LEFT **Galileo's observations of the phases of Venus and moons of Jupiter gave such powerful confirmation of Copernicus's revelation that the Earth is not at the centre of the Universe but circles the Sun, that he was brought before a papal court. Under severe pressure, he finally refuted his own ideas.**

...while its southern
hemisphere is
heavily cratered

WET RED PLANET

In 1938 the Martians were on their way to invade Earth. Orson Welles' realistic radio adaptation of *The War of the Worlds* terrified thousands of New Jersey folk into believing the Earth was under interplanetary siege. It was quite understandable: ever since the nineteenth-century astronomer Schiaparelli's observations of channels on Mars had been mistranslated as 'canals' the idea that life existed on a habitable planet had taken root in the common psyche.

In the early twentieth century Percival Lowell, an American astronomer, had speculated that the channels were a network of aqueducts created by an alien race to shift precious scarce water about the desiccating planet. He was quite wrong about the intelligent life, for the 1971 Mariner 9 fly-by saw not only no civilizations but revealed that the channels were huge natural canyons. They were not carved by water, but gashes caused by solid crust buckling outward under massive stresses. But Lowell did get it right about the loss of water.

Since the Viking probe landed on Mars in 1974 and discovered that the planet is a dry husk, there has been a puzzle about Mars. Currently its atmosphere is too thin and cold for liquid water to exist on its surface. If sunlight were to warm a Martian polar ice cap, the ice would not turn into water but simply evaporate, for there is not enough surface pressure to keep the molecules from flying apart to form a gas.

Yet as the surface images collected by the Mariner fly-bys were added to by later missions, the evidence became irrefutable: rivers once shaped the Martian surface. The 1997 Mars Surveyor images even showed depressions indicating that surface water stood in lakes for long enough to affect the geography of the planet – and maybe long enough for primitive life to sink its claws in.

Not only does the evidence suggest rivers on the planet, but features of the outflow channels show that huge volumes of water once ran across the surface in catastrophic floods. These shifted tonnes of rock across the planet, creating a smorgasbord sample for the Mars Pathfinder Sojourner rover to investigate when it landed at an alluvial delta in 1997.

Mars had been a wet red planet billions of years ago – but why? The water should have turned directly into gas, and we cannot see any evidence for a large volume of water today; currently the planet's atmosphere contains only 0.03 per cent water, a tiny fraction.

It now appears that, four billion years ago, Mars was a comfortable, warm, hospitable planet with a thick atmosphere. It is thought that its high concentrations of carbon dioxide allowed high-altitude 'dry-ice' clouds to form. These clouds caused an intense greenhouse effect, allowing solar energy in but trapping the infrared radiation generated by the surface as it heated up. Liquid water would have been stable on the surface beneath this thick atmosphere, and capable of creating the alluvial features still visible today.

Unfortunately Mars is just too small to hold on to such a thick atmosphere. As gas molecules are hit by ultraviolet light in the upper atmosphere of the planet, some dissociate into ions, and these possess enough energy to escape the Martian gravitational field. Water vapour in the upper atmosphere would have dissociated in this way, creating hydrogen ions which then escaped into space. The planet's atmosphere seeped away over a few million years, and with it the majority of its water and insulation. As a consequence, the planet slowly became the frozen dried husk we see today.

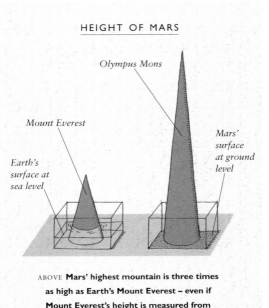

HEIGHT OF MARS

Olympus Mons

Mount Everest

Earth's
surface at
sea level

Mars'
surface
at ground
level

ABOVE **Mars' highest mountain is three times as high as Earth's Mount Everest – even if Mount Everest's height is measured from the lowest point on the Earth's surface at the bottom of the Pacific Ocean.**

WHERE DID THE MOON COME FROM?

Up until the 1970s there was an almost complete lack of consensus between geologists and astronomers as to the origin of the Moon. There were many competing theories. The first, the fission theory, required the Earth to have been spinning so violently that the tremendous centrifugal forces flung the material of the Moon outwards. Another popular idea was that the Moon was a wandering nomad in the early Solar System and was captured by Earth's gravity. Another, the co-accretion theory, had the Moon forming from the solar nebula at the same time as the Earth. None of these fitted very well with the facts as we know them, though.

The Earth simply could not have spun fast enough for the fission theory to work. If the angular momentum from the Moon was put back into the Earth, the combined planet would have a day some eight hours long – but to fling itself apart the Earth would need to rotate at least once every two hours.

The Apollo missions from 1969 to 1972 returned information that fuelled a new argument. Geological results from rock samples indicated that the Moon and Earth did indeed come from a common reservoir of materials. The samples contained isotopes of certain elements that could be used as markers to determine the common origins of material. The heavier isotopes of the elements possess a different number of neutrons than their commoner cousins, and the difference in mass can separate them in physical processes involving gravity. Although the material that formed the Earth and Moon may have come from the same proto-planetary nebula, if it took drastically different routes to get here, this would betray itself in different ratios of the isotopes to elements.

The rock samples returned by the Apollo missions showed geologists an oxygen isotope composition that was distinctively similar to Earth's rocks: the chemical signatures matched, and so they had a close common ancestry. This

ABOVE There remains considerable confusion over where the Moon came from. Could it have split off from the Earth or been captured by it?

knocked the capture theory out of contention. The Apollo results also showed that the Moon was highly depleted in iron compared with Earth, denting the fission theory. In fact the composition of the Moon had more in common with Earth's crustal layer than the planet's overall composition, iron core included, so could it be that the Moon was made from part of the Earth's crust?

Post-Apollo a new, more violent theory came to the fore in scientific debate: the big splash theory. It explains how the material came from the Earth and yet is so poor in metals. Some time in the past a huge planetoid is thought to have smashed into Earth, blowing one hundred billion billion tonnes of material out of the Earth's metal-poor mantle (the 2,900km-deep region encasing the iron-rich core). The resulting storm of debris swarming around the Earth is thought to have coalesced to form the Moon.

In 1997 researchers put the big splash theory to the test in a computer simulation, and found to their surprise that they needed an impactor three times the size of Mars to eject enough material high enough to form the Moon. This collision would need to be so violent as to relegate other asteroid impacts in Earth's history to also rans – dwarfing the collision accused of wiping out the dinosaurs. So the Moon may be a reminder of an event that nearly destroyed our planet.

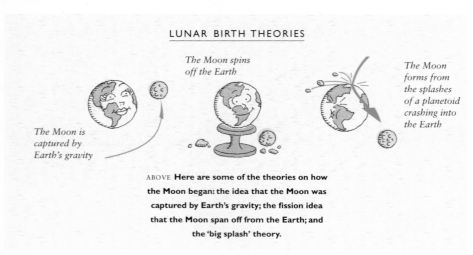

LUNAR BIRTH THEORIES

The Moon spins off the Earth

The Moon is captured by Earth's gravity

The Moon forms from the splashes of a planetoid crashing into the Earth

ABOVE Here are some of the theories on how the Moon began: the idea that the Moon was captured by Earth's gravity; the fission idea that the Moon span off from the Earth; and the 'big splash' theory.

ABOVE **Some people are absolutely convinced that aliens exist. This is the replica of the head and torso of an alien autopsied after a UFO was said to have crashed in a field in Roswell, USA, in 1947.**

RIGHT **Drawings of aliens that people claim to have seen are often remarkably similar, with large hairless heads, giant eyes and small almost foetus-like bodies.**

PERCY LOWELL
1855–1916
American astronomer Lowell saw markings on the surface of Mars which he was convinced were canals built by Martians.

LUCK OR JUDGEMENT?

The last planet to betray its existence to the ever-increasing number of sky-surveying telescopes was Pluto, named after the god of the underworld. It had been noted that there was still something unaccounted for in the motions of Uranus and Neptune, suggesting that there was yet another planet to be discovered. However the mathematicians couldn't use Neptune's motion to locate it, because Neptune moves so slowly – from our perspective – that the distance it had moved in the time it had been observed was too small to provide adequate data. Uranus had moved further, but the irregularities in its orbit were minute.

Percival Lowell, famous for speculating about intelligent life on Mars, was part of a group of astronomers who tackled the problem in the early twentieth century. From Uranus's slight perturbations he calculated two possible locations for the planet, and also its mass – around 6.6 Earth masses. Yet both of his calculations turned out to be completely wrong, and Lowell forlornly scanned the sky for the planet from 1906 until his death in 1916.

In February 1930, though, an astronomer called Clyde Tombaugh compared two photographs made by a telescope donated to the Lowell observatory by Lowell's brother, and there was a wandering pinpoint of light within six degrees of Lowell's prediction. Pluto was found.

Despite this apparent vindication, the position of Pluto within the region of Lowell's prediction proved to be coincidental. Pluto turned out to have less than a Moon's mass, and could not possibly exert any measurable gravitational influence on the orbits of Uranus and Neptune. To this day no 'Planet X' has been discovered to account for the supposed perturbations, and they may have originated as errors in early observations. But even if Lowell's work was based on flawed data, it was he who laid the path to Pluto's eventual discovery by Clyde Tombaugh.

SPECULATION AND FRUSTRATION

At the beginning of this century Lowell was one of the promoters of another type of scientific speculation about the planets: how old were they, what were they made of and what – or who – lived on them?

The extra-terrestrial life discussion began in 1877 when the Italian astronomer Giovanni Schiaparelli revealed the discovery of faint straight lines on Mars that he termed canali. He meant 'channels', but the word was translated as 'canals' with the implication that a sentient life form had dug them out. Throughout his career Lowell championed this idea, convincingly promoting the theory that an intelligent Martian species had constructed massive canals to transport water from visible polar caps of what was assumed to be ice, in a futile effort to combat the climatic deterioration of an increasingly hostile planet.

Less sensational but important questions about the planets concerned their origins, composition and surface conditions. Telescope observation suggested that the Moon, Mars and Mercury were the only bodies with mountains and dark cratered maria or 'seas' that could be oceans of ancient volcanic lava. Telescopes could not physically touch the Moon or planets, however, and we could not even tell if the surface of our closest neighbour was solid rock or a series of mountain-ringed oceans of fine dust that might engulf a landing spacecraft. The low resolutions of Earthbound telescopes and the narrow range of frequencies that penetrate Earth's atmosphere made such detailed observation impossible. What was needed was a way of overcoming the limitations of observation from Earth. The arrival of another technology led to another burst of discoveries.

ABOVE **Earth's changing surface means very few meteor impact craters survive for long. This is one of the biggest – Winslow meteor crater in Arizona, USA.**

CRATER DATA

The Russian Lunar 2, launched in 1959, is recorded as the first successful planetary exploration mission of the Space Age. It was the start of over 40 Russian and American science missions to the Moon, imaging its permanently hidden far side, crashing – or 'hard landing' as the euphemism goes – into its surface, running rovers over it and culminating in the Apollo manned landings. Every successful mission has enhanced our understanding of our nearest neighbour in space, and has helped to unravel a few planetary puzzles as well.

Some of the Moon's more mysterious features were clearly visible from Earth: its craters. Originally they were thought to be volcanic in nature, like the majority of obvious craters here on Earth. As visible impact craters are rare on Earth, most lunar geologists reasoned that they should not be a major feature of the Moon.

BELOW **Unprotected by an atmosphere and unchanging geologically, the Moon's surface bears the scars of thousands of meteor impacts.**

The geologists were not unanimous in this opinion, however. G. K. Gilbert, a scientist with the US Geological Survey in the 1890s, drew attention to the fact that the Moon's large craters – mountain-rimmed circular features with floors that often lie below the surrounding plains – are quite different to the Earth's volcanic craters. These tend to be smaller, deeper and almost always occur at the tops of volcanic cones. Gilbert believed that this meant the lunar craters were not volcanic, and were in fact impact craters. But there was an objection.

There are no elliptical craters on the Moon, and this circular symmetry posed a problem. If you throw stones into a sandpit they form elliptical pits, not round ones, because they strike the sand obliquely. Asteroids and similar bodies also normally strike at oblique angles. The solution to this problem was reached through comparative studies of bomb and shell craters, which are always roughly circular regardless of the impact angle of the projectile. Such a crater is created not by the impact, but by the subsequent explosion.

The gravitational pull of the Moon always accelerates impacting bodies to at least 2.4 km per second before they hit the lunar surface. The resultant impact is packed with so much kinetic energy that the materials explode. So the Moon's craters are not so much impact as explosion craters, and this removed the main objection to the impact theory. Once the Moon's craters were shown to be created by exploding impacts, they provided an important clue to the age of features on other planets in the Solar System.

On any moon or planet, heavily cratered terrain will always be older than surfaces with fewer craters because it has been exposed to impact for longer. The cratering rate throughout the inner Solar System should be broadly the same at any given time, and reflect the number of small bodies flying around the Solar System at that time. If we could calibrate the relationship between crater numbers and the age of the terrain, we would have a useful tool for estimating the ages of surfaces on other cratered bodies like Mercury and Mars.

The surface of the Moon is covered in craters

Moon craters are geologically different from those on Earth, shallow, large and ringed with mountains

WATER ON THE MOON

When the final Apollo mission left the Moon in 1972 we knew it was a barren world composed of little more than silicate rock, similar in composition to the Earth but depleted of iron and other metals. The Moon contained no volatile molecules either – no water or gases – for its limited gravity had no hope of clinging onto light molecules as its surface baked to over boiling point every lunar day.

If it were not for its lack of volatiles the Moon might make a useful base for exploring the rest of the Solar System. Currently every probe, space station module and astronaut must be pushed out of the gravitational well of the Earth on expensive launchers. Not only that, but all the consumables and, most importantly, all the fuel for a mission – often half the mass – must be hauled into orbit too. The Moon would be a much better place to assemble such missions, perhaps even manufacturing components and mining fuel on the surface at a fraction of the cost. The gravity, one-sixteenth of Earth's, is strong enough to make working on the Moon comfortable but not strong enough to make launching from the Moon into deep space expensive. But with no water on the Moon the plan falls down, because we would need to ship a huge amount of water to the Moon for life support, manufacturing and fuel production.

After Apollo 17 left the Moon we did not return until 1994 when the US military used it as target practice for the 'Star Wars' Strategic Defense Initiative anti-ballistic missile programme. The Clementine mission's objective was to demonstrate anti-missile technology by orbiting the Moon testing new sensor technology, and then targeting an asteroid on a fly-by – a test mission with surprising similarities to tracking and destroying an enemy ballistic missile.

Clementine, orbiting the Moon at only a 100km above the poles, returned the most spectacular high-resolution pictures of the lunar surface ever seen. The satellite took many small strips of pictures as it swept over the Moon's surface every two hours; within a month the Moon had fully spun on its axis

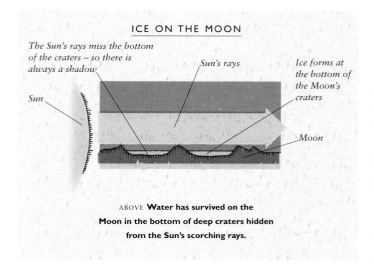

ICE ON THE MOON

The Sun's rays miss the bottom of the craters – so there is always a shadow

Sun

Sun's rays

Ice forms at the bottom of the Moon's craters

Moon

ABOVE **Water has survived on the Moon in the bottom of deep craters hidden from the Sun's scorching rays.**

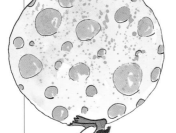

LEFT **In 1998 NASA's Lunar Prospector** space mission revealed that there is water on the surface of the Moon, frozen in shady craters near the poles. Perhaps the next mission will find penguins…

beneath Clementine's orbit, providing enough pictures to create a global mosaic of its entire surface.

Clementine, the lunar south pole and the Deep Space Network dishes were lined up for the experiment during four orbits in early April 1994, but the experiment worked on only two of these orbits. The results were tantalizing. On Clementine lunar orbit 234 the reflected signal showed just the right signature, indicating that there might be water on the Moon.

The findings were ambiguous at best, but the prospect of water ice on the Moon was too exciting to ignore: if it did exist there, then humankind could use it to explore the Solar System. Water can have obvious uses in life support systems, but split it into its constituent hydrogen and oxygen and you have one of the most potent rocket fuels known.

Hydrogen can also be used to remove oxygen from the numerous oxides on the Moon's surface, to manufacture pure metals for construction work or shipping back to Earth. Water would turn a lonely companion into a well-stocked stepping-stone to the stars.

In 1998, NASA's Lunar Prospector mission was sent to confirm this. Lunar Prospector was to take an inventory of the materials on the Moon and scan the surface for hydrogen, always a tell-tale sign of water.

In March 1998 NASA announced that the Moon did indeed exhibit a signal indicating that water was hidden – not only in the deep depression at the south pole, but also in shallower craters at the north pole. So there really is water on the Moon.

This has proved possible using calculations based on the number of craters on the lunar maria and the number of projectiles currently whizzing around the Solar System, backed up by radiometric dating of Apollo samples from the maria. The results indicate that for a body the size of the Earth, a crater with a diameter of one kilometre should occur every few ten thousand years, a couple of 10km craters every million years, and a 100km crater every 50 million years. The rates indicate that the lunar maria have been pounded at their current rate for several billion years, and this agrees with radioactive dating which indicate them to be 3.3 to 3.8 billion years old.

The lunar highlands raised a problem, though. In order to accumulate their current number of craters they would need to have been exposed to bombardment at the current rate for 38 billion years – which is much longer than the age of the Universe. As this is impossible, there must have been a period of much greater bombardment early in the history of the Solar System. This is in agreement with images of the satellites of Jupiter and Saturn, which show surfaces as heavily pitted as the lunar highlands.

The heavy bombardment must have occurred within the first billion years of the Solar System. This means that the technique can be used to

roughly date old cratered features formed in the early days of the Solar System, and more accurately date newer features formed during the period of the current cratering rate. So planetary scientists looking at pictures from the Galileo and Cassini space missions showing the satellites of Jupiter and Saturn are able to give rough dates for their features, despite having no physical samples.

ENIGMATIC VENUS

Clues from the Moon enabled astronomers to get a virtual foothold on planets further afield, such as Mars and Mercury. Venus, however, was still a complete mystery until the 1960s, for even though Venus comes closer to Earth than any other planet, its opaque atmosphere makes its surface impossible to observe using ground-based telescopes. The answer lay in the new space technology, and in December 1962 Mariner 2 became the first successful probe to fly-by Venus and send back information about its nature.

The data proved a surprise, indicating that the surface temperature was at least 380°C. Astronomers had done calculations showing that a planet of Venus' reflectivity and distance from the Sun should have had a surface temperature of about –20°C. So why was the surface so much hotter than the calculations would suggest? The cloud tops of the planet were at about the calculated temperature, but beneath this was – according to later surface landing probes – a dry pressure-cooker of an atmosphere. The later Russian Venera probes showed that the surface pressure was 90 times that on Earth – roughly the pressure found at ocean depths of 1,000 metres – and found that the temperatures obtained by Mariner 2 were about right.

The late Carl Sagan, then a graduate student at the University of Chicago, conducted calculations establishing that the high temperatures resulted from an extreme example of the greenhouse effect, in which the atmosphere acts as a massive

ABOVE **Mercury is the nearest planet to the Sun, scorching by day at over 430°C and yet freezing by night at -180°C – because it has virtually no atmosphere to keep it warm.**

LEFT **Our nearest neighbour, Venus, is shrouded in a very thick atmosphere, rich in carbon dioxide – which keeps the surface temperature at a scorching 480°C. Here radar has stripped the cloud cover to reveal the planet's surface. The 'hole' at the top of the planet is the gap in NASA's map.**

The further away a galaxy is, the faster it moves away from us

Splashes of paint on the balloon represent galaxies in space

insulating blanket for the planet. When sunlight penetrates the Venusian atmosphere it is absorbed by the surface and re-radiated as infra-red radiation. But since the atmosphere is 97 per cent carbon dioxide – a gas transparent to visible sunlight, but opaque to infra-red radiation – the radiation gets in but cannot get back out. Not only did the effect increase the temperature of the planet, but it also appears to have liberated more carbon dioxide from rocks and oceans.

The evidence for the past existence of oceans lies in the ratio of heavy to light hydrogen isotopes in the atmosphere. Water vapour tends to dissociate into hydrogen and oxygen when exposed to solar ultraviolet radiation, as in the upper atmosphere of Venus. Hydrogen, being the lightest of the elements, can escape from a small planet's gravitational pull, but the lightest isotope – normal hydrogen – is lost more readily than deuterium, the more massive isotope. This naturally increases the ratio of deuterium to normal hydrogen. As the NASA Pioneer Venus probe descended through the Venusian atmosphere in 1978, Thomas Donahue of the University of Michigan – a champion of the idea of Venusian oceans – saw results indicating a deuterium-hydrogen ratio about a hundred times that of the Earth. This showed that enough water for modest-sized oceans had been lost from the planet.

TO JUPITER AND BEYOND

Up until 1972 the outer Solar System had been ignored by the space agencies, but in that year Pioneer 10 was launched by NASA as a pathfinder to the gas giants: Jupiter, Saturn, Uranus and Neptune.

No-one quite knew what to expect out there. No spacecraft had ever travelled further than the orbit of Mars, beyond which astronomers had seen a thick belt of rocky remnants from the formation of the Solar System, known as the Asteroid Belt. Any missions to Jupiter or beyond would have to traverse it. The risk it posed to spacecraft was unknown. It was speculated that the belt might be packed with spacecraft-killing micro-meteoroids, effectively making it a 280 million-kilometre-deep barrier blocking planetary astronomy's exploration of the outer Solar System. Mission managers knew nothing about the distance between the particles or how they were distributed in size, and a destructive collision seemed very likely. As it turned out Pioneer 10 made it through the danger zone without a scar, and so far we have never lost a probe to the Asteroid Belt. It is probably composed of a few large chunks rather than many small pieces.

In December 1973 Pioneer 10 became the first spacecraft to encounter Jupiter, revealing details about its vast magnetosphere and the ferocious radiation belts surrounding the planet (1,000 times greater than the spacecraft-killing belts around the Earth). Only the spacecraft's speed saved the electronics aboard from being fried by plasma radiation – charged electrons and ions controlled by the intense magnetic fields of Jupiter's magnetosphere.

ABOVE **Ever since the Big Bang which gave birth to the Universe, the galaxies have been flying further and further apart as space expands. The increasing separation of galaxies as space stretches can be imagined by drawing them on a balloon, and watching them move further and further apart as the balloon is inflated. Every galaxy seems to be at the centre of the expansion.**

Many planets have their own strong magnetic fields, and these form comet-shaped cavities around the planets, dragged out by the solar wind. The cavities, or magnetospheres, control the movement of basic charged particles within, trapping them from the solar wind and accelerating them to form rings of radiation girdling a planet. Even from the distance of Earth, Jupiter crackled on the radio spectrum revealing the radio signature of high-energy particles close by – a tell-tale sign that it may be one of the most hazardous places in the system.

Pioneer 10's success in penetrating the radiation belts, and some fortunate planetary alignments, prompted a decision to alter the trajectory of its sister craft Pioneer 11. Shortly before its launch in 1974 Pioneer 11 was deliberately targeted closer to Jupiter than originally planned, so the planet would throw it out in a new direction towards Saturn – a technique the later Voyager missions were to use to go on a 'Grand Tour' of the outer Solar System. But the radiation belts were not to be ignored so lightly, for during its closest approach to Jupiter the spacecraft missed hoped-for images of Jupiter's moon Io because of radiation-induced rogue instructions.

THE GRAND TOUR

While Pioneer 11 was still on its five-year voyage to Saturn, two new spacecraft were despatched to the outer Solar System. Voyagers 1 and 2 were launched from the Kennedy Space Center, Florida, in August 1977. The Voyager missions were designed to take advantage of a rare geometric arrangement of the outer planets which permitted a four-planet tour in the shortest possible time, without the need for large rocket motors. Each planetary encounter lent energy to the spacecraft, and bent its flightpath enough to propel it to its next destination. This 'gravity assist' technique reduced the flight time to Neptune from 30 years to 12.

The Voyagers have an overwhelming array of findings to their names. Their cameras returned stunning images of Jupiter, showing storms in its churning atmosphere for the first time. They also discovered an unexpected feature missed by the Pioneers, for Jupiter had a barely discernible ring, impossible to see from Earth. The spacecraft provided the first photos of the major Jovian moons – pictures that are only now being bettered by the Galileo probe.

VOLCANOES AND ICE OCEANS

Evidence of active vulcanism elsewhere in the Solar System was revealed when Voyager caught a glimpse of a volcano erupting at the edge of Io's disc: a plume 30 times higher than Everest showering an area of the size of France with lava. The volcanoes on Io are now thought to be the primary source of Jupiter's radiation belts, pumping them up along with high-energy particles of sulphur and oxygen.

Voyager discovered the permanent ice-oceans of Europa which possibly have liquid oceans underneath, and the curious mixture of old and new crust on Ganymede. It returned images of Callisto, which has craters standing shoulder-to-shoulder on its ancient surface, and subsequently went on to discover the three new moons Adrastea, Metis and Thebe, all considerably less than 100km in diameter.

Voyager showed shepherd moons herding the rings

The Voyager missions revealed much about Saturn's rings and moons

Saturn's rings are thin discs of tiny lumps of ice

RINGS OF ICE

Shortly afterwards Pioneer 11 was providing the first close-up images of Saturn, and important information required to navigate the two Voyagers through the complex system of moons and rings surrounding Saturn. The Voyagers took a more direct route to the ringed planet, and got there a year later. The fantastic array of images they returned revealed Saturn's rings to be far more complex than expected, with many thousands of ringlets within each major ring, each composed of trillions of ice particles and car-sized icebergs. Some had curious spokes and some were kinked in defiance of the laws of celestial mechanics. No-one knew how they remained stable until close observation of Voyager's pictures revealed tiny shepherd moons on the rims of the rings herding their flocks of ice within the ring systems, counteracting the tidal forces that would otherwise rip the rings apart.

Voyager 1 skimmed past the moon Titan, finding that it was too cold for life but that the atmosphere is like that of a young Earth. The rendezvous with Titan meant that even the gravitational forces of giant Saturn could not bend its path to follow Voyager 2 on its course to Uranus and Neptune. Meanwhile the moons Atlas, Prometheus, Pandora, Epimethus, Janus, Helene, Telesto and Calypso were discovered, as well as the Saturnian auroral lights. Voyager 1 then had another critical encounter with Titan. Mission control attached a great deal of importance to this objective.

THE CHILDREN OF THE OTHER STARS

While we wait for technology to advance to the point where we can actually see planets, the basic principles of physics have ensured that we are not blind to their presence. Newton's second law states that for every action there is an equal and opposite reaction, so if a star exerts a pull on a planet, so a planet must exert a pull on the star. Just as you do if you take a bucket of water and spin around with it like a top, the two bodies spin around a common centre of gravity and the star appears to wobble back and forth slightly. The period of the wobble reveals the time it takes for the planet to orbit the star.

Yet that is still not enough to be detectable. The wobbles are far smaller than the orbits of the planets we are looking for, and indistinguishable to direct measurement. However, in 1842 Christian Doppler pointed out that if a light source is approaching or receding from the observer, the light waves will be crowded together or spread out. This is the Doppler effect – the same effect that makes an ambulance siren change tone as it approaches and then passes. And it so happens that, as the unseen planet pulls the star back and forth, we do just about have the spectral resolution needed to see the light from the star shift up and down the spectrum.

We need to take account, not only of the wobble in the light-spectra caused by the motion of the planet about the star, but also the Doppler shift caused by the motion of the Earth about the Sun, the Moon about the Earth and even the planets near Earth. Once this is done, current telescopes can detect the motion of stars to within three metres per second – enough to see the wobbles caused by alien Jupiters.

The period of the wobble is equal to the time the planet takes to orbit its star. From Kepler's third law (yet another basic law in our ancient physics toolkit) the planet's average distance to the star is equal to the cube root of the square of the period – so in turn we know the planet's orbital distance. The amplitude of the wobble – its size from peak to trough – tells us something of the mass of the planet, or at any rate the minimum proportion of the masses of the planet and star. The larger the amplitude the larger the planet must be with respect to the Sun. It is a minimum figure because we cannot tell whether the planet is in orbit exactly edge-on to us (maximizing the apparent wobble effect) or somewhere between that and at right angles to us. From the brightness of the star and our stellar models we can work out its mass, and so get a minimum mass for the planet from the fraction.

The first planet to be seen using the Doppler technique was an extra-solar planet orbiting the star 51 Pegasi by Michael Mayor and Didier Queloz of the Geneva Observatory. Since then seven others have been detected. They are not Earth-like, for they are massive Jupiter-sized bodies orbiting incineratingly close to their stars, but these planets are the ones we expect to see first using the developing optical technology. As that technology is refined, though, it may not be long before alien Earths are seen as children of other stars.

URANUS AND NEPTUNE

NASA decided that if Voyager 1 were to miss Titan, Voyager 2 would be re-tasked to visit the moon, forsaking its mission to Uranus and Neptune. Since Titan is one of the few satellites large enough to have its own atmosphere, researchers were anxious to find out as much about it as possible. In the event, although Titan was too cloudy to see the surface all the experiments were successful, allowing Voyager 2 to target Uranus.

Voyager 2's fly-past of Uranus, the third largest planet in the Solar System, confirmed that the planet was tipped on its side – an unusual position thought to be due to a collision with a planet-sized body early in the Solar System's history. Because of its odd orientation, with its polar regions in sunlight or darkness for half a Uranian year, scientists did not know what to expect. The presence of a magnetic field around Uranus was unknown until Voyager encountered it, and researchers found a remarkable magnetosphere, twisted by the planet's rotation into a long corkscrew shape behind the planet.

The radiation belts were found to be sufficiently large and intense to irradiate any methane trapped in the icy surfaces of the inner moons and rings, and this may explain the uniformly grey colour of those bodies. Voyager found a pair of such moons, each less than 150km across, and its pictures of Miranda, the innermost large moon, show it to be one of the most peculiar bodies in the Solar System. The moon appears to have huge faults 20km deep and a mixture of old and new surfaces, indicating that Miranda may be a moon in the process of coming together after a violent impact blew it apart.

Three years later Voyager 2 became the only spacecraft that has ever encountered Neptune, and the visit solved many questions about Neptune's rings. From Earth the rings do not appear

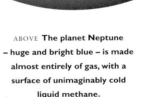

ABOVE **The planet Neptune – huge and bright blue – is made almost entirely of gas, with a surface of unimaginably cold liquid methane.**

complete, but appear as ring-arcs or partial rings. The puzzle was solved by images that showed that the rings are actually complete, but composed of such fine material that they cannot be fully resolved at Earth. Voyager also monitored a hurricane-like 'Great Dark Spot' in the planet's atmosphere, reminiscent of Jupiter's Great Red Spot. Originally Neptune was thought too cold to support major atmospheric disturbances such as large-scale storms, since it receives radiation from the Sun at only 0.1 per cent of the strength at Earth's orbit. For the first time the aurorae of Neptune were seen and six new moons in total were discovered – all as 'dark as soot', reflecting only six per cent of the almost insignificant amount of light that hits them.

Voyager 2 finished its planetary career in 1989 by skimming Neptune's moon, Triton, and seeing geysers spewing nitrogen gas and black particles several kilometres into the moon's tenuous atmosphere. Triton seemed to be a dense body, and its retrograde, or backwards, orbit has led researchers to believe it was not part of Neptune's original family, but a captured planetoid.

A COMPLETED MODEL

Robot spacecraft are continuing to explore the strange worlds surrounding us. However, there is now a limited supply of secrets left to unravel in the Solar System. In 1983 IRAS, an infra-red orbiting astronomy mission, eliminated the possibility of finding any further planets. By scanning all the sky in infra-red twice, and comparing the images, every moving object was identified. No further wondering stars were discovered, and so the Solar System has been confirmed in its complement of nine plants. The work begun by Copernicus and Galileo was complete.

AT A GLANCE

EARLY TELESCOPE

AD 140

Ptolemy devises a complete model of the motion of the heavens based on orbits and small sub-orbits called epicycles which explains the way planets occasionally perform strange loops.

NICOLAUS COPERNICUS

543

Nicolaus Copernicus shows that the Earth is not at the centre of the Universe and is not even fixed but is rotating rapidly about its own axis.

1610

Galileo Galilei uses the telescope to show that the Milky Way is composed of distant stars and that Venus has phases just like the moon – thus proving Copernicus' theory.

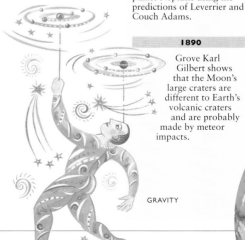

GRAVITY

1781

William Herschel discovers the planet Uranus, the first new planet to be discovered since ancient times.

HERMANN HELMHOLTZ

1843

John Couch Adams analyzes perturbations of Uranus's orbit and predicts the position of an undiscovered planet within two degrees. His calculations are repeated by Urbain Leverrier two years later.

ASTEROID

1846

Johann Galle discovers the planet Neptune using the predictions of Leverrier and Couch Adams.

1890

Grove Karl Gilbert shows that the Moon's large craters are different to Earth's volcanic craters and are probably made by meteor impacts.

1930

Clyde Tombaugh discovers the planet Pluto by comparing two photographs made by a telescope donated to the Lowell Observatory.

1959

The Russian Lunar II probe is the first successful space mission to another planet, circling the moon and revealing its dark side for the first time.

1962

The Mariner II probe flies by Venus and shows that the surface of the planet is at a temperature of 380°C – 400°C hotter than expected.

1964

Carl Sagan realizes that the high temperatures on the surface of Venus are due to a greenhouse effect, created by the fact that its atmosphere is 97% carbon dioxide.

1969

The Apollo missions bring back rocks from the Moon reveals that the Moon's rocks are formed at the same time as the Earth's.

THE EXPANDING UNIVERSE

1973

The Pioneer 10 probe becomes the first space probe to encounter Jupiter and reveals a great deal of detail about its magnetosphere and radiation belts.

1974

The Pioneer 11 space probe provids the first close up images of Saturn and a great deal of invaluable information about its moons and rings.

1980

Voyager I goes past Saturn's moon Titan and reveals that although it is very cold it has an atmosphere like that of the early Earth. It also discovers eight new moons of Saturn.

1986

Voyager II flies past Uranus and shows the planet is tipped on its side. It also discovers ten new moons of Uranus.

1989

Voyager II becomes the first space craft to encounter Neptune and also observes geysers spewing nitrogen gas on the surface of Neptune's moon Triton.

1990

The Hubble Space Telescope is launched and once its optics are corrected it provids views of the Universe of startling clarity.

1994

The American Clementine mission provides the most spectacular high resolution pictures of the Moon's surface ever seen, and provides enough pictures to create a detailed 3D map of the entire surface.

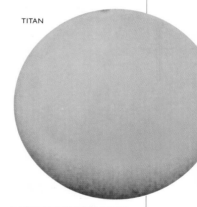

TITAN

1997

The Mars Pathfinder Sojourner lands on Mars and sends back extraordinary live pictures.

1998

The Lunar Prospector discovers water on the Moon as ice beneath the surface.

WATER ON THE MOON

VENUS

The Nature of Matter

CHEMISTRY

In the laboratory of the future, the chemist will not be wearing a white lab coat. You will not find her juggling test tubes, but sitting in front of a zero-flicker polymer LCD screen, maybe 60 centimetres across. On the screen, the image of a complex drug molecule. The structure rotates slowly as, using a double blink of the eye, she manipulates the image, tweaks a chemical group here and alters a bond there.

Another blink, this time towards the lower left edge of the screen. The molecule's compressed data file is uploaded quickly to the Supernet – the millennial successor to the grindingly slow Internet – and in less than two seconds a collaborator on the other side of the world sees the new drug on his computer screen too. The two scientists chat briefly via a high-speed audio-video link and our chemist double-blinks again on the icon of a small reaction flask at the top left of the screen. On the other side of the continent, in an underground factory, a robot arm twitches to life and begins to mix ingredients.

Science fiction? Perhaps, but in the relatively few years since we began to manipulate substances for our purposes the art of molecular architecture has come a long way. The main function of chemistry will always be the creation of substances in reaction flasks of some kind which we can use for medicine, for materials, for agriculture and even – paradoxically to some – to cure environmental ills. But how we discover new compounds, how we manipulate them and how we synthesize new ones will follow new routes as technology expands and develops in the future.

It will not be too long before chemists can make the compounds we need for the lifestyle that we have become accustomed to with minimal risk, zero solvent waste and 100 per cent atom efficiency.

No molecules of starting materials will be wasted, as each will make one molecule of product with nothing left over. Molecules may even be built up piece by piece from individual atoms. And it is likely that chemists will be able to do all this without having to set foot in a traditional laboratory, handle a test tube or don a white lab coat ever again.

BELOW **The science of chemistry has come a long way from the days when white-coated technicians played around in the laboratory with flasks of mysterious, steaming, bubbling liquids – and frequently blew themselves up. Now most cutting edge chemistry is conducted in a virtual world on computers.**

ABOVE **The Ancient Greek philosopher Aristotle viewed the world's substances in terms of four different elements – earth, wind, fire and water. This idea of the four elements persisted for almost two thousand years and formed the basis for the alchemists' view of the material world.**

TRIAL AND ERROR

Chemistry is not new. The people of the ancient world did chemistry: they extracted ores from the earth to make tools and weapons, they fermented vegetable matter to make alcohol – admittedly more biotechnology than chemistry – and attempted to understand the behaviour of the substances they used. The four Aristotelian classical elements – earth, wind, fire and water – served as a reasonable basis for them to classify the materials around them, and the Ancient Greek philosopher Democritus hit on the idea of atoms to explain what things were made of.

A kind of science, drenched in superstition, sprang from this basic and ancient understanding: alchemy. It had some rather dubious aims. The most notorious was the attempt to convert one substance into another by transmutation of some kind, most of the effort being directed at creating gold from base metals such as lead. The alchemists even believed in the possibility of immortality: find the right chemical cocktail and the human body could be preserved alive and intact forever.

The alchemists were not short of 'proof' for such possibilities. There was, for instance, the existence of various substances that could spontaneously smoulder and burst into flames. Many of these were later discovered to be phosphorus compounds. There was quicksilver – the liquid metal mercury – an obvious hint of mystical happenings in the world around us. Numerous other inexplicable chemical phenomena provided the alchemists with all the proof they needed that they were on the right path to riches and immortality.

The extremes to which the alchemists went in their pseudo-scientific efforts are sometimes hard to believe. In the middle of the eighteenth century the German merchant Henning Brand, who enjoyed alchemy as a hobby, discovered an eerie glowing element in the course of a bizarre experiment connected with turning lead into gold. For some reason known only to him he acquired 50 to 60 buckets of urine and left them until they began to putrefy. He then boiled the urine for two weeks or so, reducing it to a paste which he then heated with sand. Whether he knew what he was doing or not is uncertain, but the resulting mixture yielded elemental phosphorus.

A SCIENTIFIC APPROACH

The modern age of chemistry really began when the base pursuit of gold and immortality gave way to a more rational approach, involving carefully devised experiments designed to test theories about the nature of matter. With this transition, chemistry emerged from its medieval miasma of mysticism and superstition and became a true science.

The Anglo-Irish scientist Robert Boyle began the process by bubbling air through various pieces of apparatus until he had proved that the air was made up of tiny particles, or what he referred to as corpuscles. In his best-selling book *The Skeptical Chymist* of 1661, Boyle attacked the Ancient Greek view of the four elements. He suggested that different substances were made, not from these classical elements, but from different primary particles that formed the corpuscles: what we now know as molecules.

Boyle took his idea only halfway to its necessary conclusion – it was after all only the seventeenth century, and analytical equipment had some way to go before it was capable of revealing the truth about the corpuscles and primary particles. He also failed to grasp the true nature of the elements; that insight would have to wait until the 1800s. But despite this Boyle's ideas were a turning point for chemistry, and his gas laws – which describe the physical behaviour of air under different pressures, temperatures and volumes – are still valid today. Almost single-handed, Boyle wrested chemistry from the clutches of the alchemists and put it on a much more solid scientific footing.

THE FOUR ELEMENTS

The Ancient Greeks believed that everything derived from four elements: earth, air, fire and water. The concept is attributed to the magician Empedocles, who was born at the beginning of the fifth century BC. According to Empedocles, at the creation of the world all four elements were mixed and held together by the force of 'love'. Then they were split up by 'strife' to form the sky, the Sun and the Earth with its oceans. Everything on Earth was built up from combinations of the four elements. Bone, for example, was reckoned to be three parts fire and two parts earth. Empedocles believed that eventually love would again triumph and the process would go into reverse, mixing the elements together again in a love-bonded whole.

Today the concept of the four elements seems simplistic, but Empedocles' ideas have found an uncanny echo in current theories about the **Big Bang** that created the Universe, and the possibility that the expansion of the Universe may go into reverse and culminate in a 'Big Crunch'. Even his four-element theory has been to some extent vindicated by the discovery that all modern elements are composed of different arrangements of the same basic particles, cooked up by fusion reactions within stars.

Earth

Air

Fire

Water

ABOVE **The Greek view of the four elements is not so very different from simple modern ideas about basic particles.**

An excess of choler brought an energetic, choleric temperament

An excess of phlegm brought about the calm and tough phlegmatic temperament

A surfeit of melancholy brought a sad and gloomy temperament

A sanguine temperament resulted from dominance of the blood

LEFT **Alchemists and astrologers viewed not only material substances but the human body in terms of the four basic elements – earth, air, fire and water. Different mixtures of these four elements were deemed to be responsible for different humors and temperaments.**

EXPERIMENTING WITH AIR

Building on the foundations laid by Boyle, eighteenth-century scientists such as Joseph Black began experimenting in earnest with the aim of uncovering the secrets of matter. Working in the 1750s, Black found that a certain type of 'air' – obtained from someone's breath, from burning or from fermentation – could combine with solid substances such as quicklime. This type of air could then be recovered from these solids simply by the action of heat. By a process of experiment Black had discovered carbon dioxide.

The existence of a constituent gas within air brought to a head the idea that 'air' was not in fact a single substance at all, and eradicated once and for all – among scientists – the notion that air was a fundamental element.

Black's work effectively pre-empted that of the more famous 'Newton of chemistry' Antoine Laurent Lavoisier, who untangled the reactions that take place when something burns. Until Lavoisier's experiments, it was thought that when something burned it gave off a 'principle' known as phlogiston: an ephemeral substance that helped to explain where the missing matter goes when, for example, a stick is burnt to ash. According to this theory it evaporated as phlogiston, never to be retrieved.

In 1772 Lavoisier reported to the Academy of Sciences that sulphur and phosphorus, when burned, actually increased in mass because they absorbed 'air'. On the other hand, the metallic lead formed when litharge (which is a red, crystalline form of lead monoxide) was heated with charcoal actually weighed less than the original litharge because, according to Lavoisier, it had lost 'air'.

Lavoisier did not at this point know the exact nature of the 'airs' involved in these processes, for Black's work was not well known at that time. Lavoisier's results, however, did demonstrate that phlogiston could not possibly exist, and that combustion involves the combination of matter with a gas. And so another mystical element had vanished into thin air.

When, in 1774, Joseph Priestley prepared 'dephlogisticated air' by heating a red precipitate of mercury, Lavoisier suspected that only a portion of atmospheric air was actually being used up in Priestley's process. He found that, during the burning process, the material absorbed this active agent. He called the active agent oxygen, a word which is derived from the French for 'origin of acidity'. The 'non-vital air' left behind was nitrogen.

ABOVE **Antoine Laurent Lavoisier (1743–94).** Lavoisier, pictured here with his wife, demolished the idea that when something burned it gave off a mysterious substance called phlogiston – and demonstrated the importance of careful measurement of substances involved in chemical reactions.

THE ELEMENTS DISCOVERED

In 1789, Lavoisier went on to publish the first proper list of the known chemical elements. Many of these, including gold, silver, copper, iron, lead, tin and mercury, were already known – even to the ancients – because they are found in natural deposits in their elemental state.

Following Lavoisier's lead, chemists such as Humphry Davy spent the next few years uncovering the constituents of many more substances. Davy, perhaps more famous for his invention of the miner's safety lamp, picked out sodium and potassium by passing electricity through solutions of sodium hydroxide and potassium hydroxide. The metals were deposited on the negative electrode in a process known as electrolysis. The development of increasingly sophisticated analytical methods allowed chemists to extract or test for elements in the most complex mixtures and compounds, and as a result more and more elements were discovered. Yet despite the apparent speed of chemical discovery prior to the twentieth century, one in four of all the elements known today were unknown before the early 1920s.

BELOW **In his hugely influential book** *Elementary Treatise on Chemistry* **(1789), illustrated with plates by his wife, Lavoisier gave his definition of a chemical element as the 'last point which analysis can reach'. In the book, he gave a working list of the chemical elements including gold, copper, silver, iron, lead, tin and mercury.**

SIR HUMPHRY DAVY
1778–1829
English chemist Humphry Davy discovered a number of important chemicals, including sodium, potassium and laughing gas, nitrous oxide.

ACIDS AND BASES

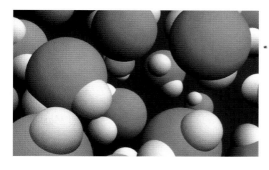

All chemicals are either acids, neutral or bases. It all comes down to their behaviour in the presence of water. In pure water, a small number of the water molecules ionize: each breaking up and losing or gaining electrons to form a positive hydrogen ion (which is a single proton) and a negative hydroxide ion. Since they were formed by splitting water molecules, the number of hydrogen and hydroxide ions is equal. But when some compounds are dissolved in water, or react to it, this neat balance is upset.

Compounds that release hydrogen ions into the water tip the balance towards acidity. Such a compound can only be described as an acid, therefore, when it is dissolved in water. Bases, on the other hand, are compounds that neutralize this imbalance by taking up the hydrogen ions in an acid – or even tipping the balance the other way, creating a surplus of hydroxide ions. An alkali is a base which dissolves in water.

JOHN DALTON
(1726–1824)

The British chemist John Dalton was the first to work out that each element is made from tiny identical particles called atoms, which is the basis of all modern chemistry.

DALTON AND ATOMIC THEORY

By the turn of the nineteenth century, chemists had begun to realize that they were dealing with a system of elements that had become too unwieldy to handle without looking a little closer at just what was meant by an element.

The experimental prowess of Boyle had discredited the mysterious and vague notions of the medieval alchemists by showing that the principles and four elements could not be combined to form all other substances. He declared that an element was a substance that could not be broken down into simpler substances. Yet what this fundamental 'stuff' was remained unknown to him. Meanwhile Lavoisier showed that there was simply no way that water could be distilled to produce earth, no matter how hard one tried, destroying another old notion.

It took British chemist John Dalton, working in the early 1800s, to figure out that each element consisted of many small, identical and indestructible particles which he termed atoms, from the Greek term coined by Democritus. His idea helped explain the difference between the fundamental substances iron, sulphur, mercury, oxygen and so on. He said each element is made of atoms of different mass, and that this mass is characteristic of the atom in question.

Dalton went onto consider what was involved in other materials in terms of atoms. He reasoned that the different atoms could combine in specific ratios to form substances that were not elements – the chemical compounds in all their diversity. Several years later, the Italian physicist Amadeo Avogadro realized that the component compounds of a gas are made up from fundamental units he called molecules, by which he meant a particle containing two or more of Dalton's atoms. The basic scheme of atoms and molecules arrived at by Dalton and Avogadro underpins all modern chemistry.

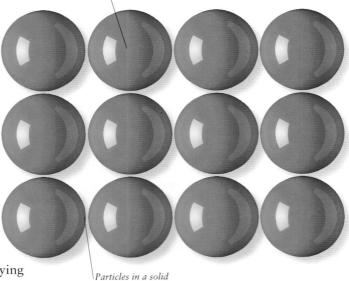

Particles in a solid are so strongly bound that they cannot move about, and therefore simply vibrate.

BELOW A solid maintains its rigid structure and shape because it is made from particles that are linked firmly together in a structure by strong bonds.

Particles in a solid are packed tightly.

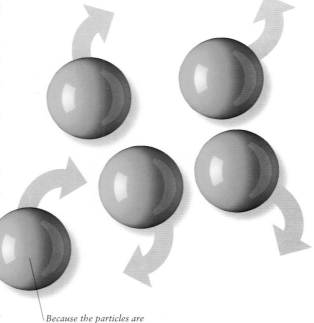

Because the particles are far apart gases can be compressed or expanded.

ABOVE Particles in gases are far apart and move very fast, and a gas spreads quickly to fill any space it is put in.

Particles in a liquid are free to move about.

ABOVE **In a liquid, particles attract each other and clump together in random clusters. However, because these clumps can slide past each other a liquid can easily change its shape, flowing to fill any container it is put in.**

AVOGADRO'S NUMBER

Avogadro not only came up with the concept of molecules: he also hit on a way of counting the number of molecules or atoms in any given amount of substance. The numbers that arose from his calculations were extraordinary, and show just how small these units are. By Avogadro's reckoning, 12 grams of carbon (graphite) contain 6×10^{23} or 600,000,000,000,000,000,000,000 atoms of carbon. This huge number is referred to as Avogadro's number, or the Avogadro constant, because it can be used to find the number of atoms in any element. This gave chemists the ability to measure and manipulate compounds far more effectively, and work out exactly what was going on in any particular chemical reaction.

RIGHT **Atoms are so small that if every atom in your body was the size of a fingernail, your body would be so big that you could hold the entire world in your hand.**

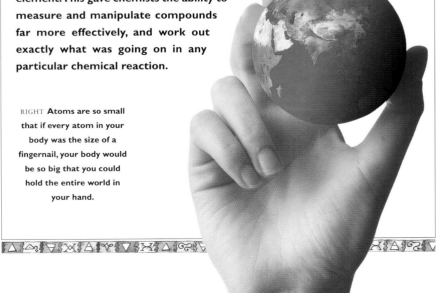

WÖHLER SNUFFS OUT THE VITAL SPARK

The understanding of atoms, elements and molecules brought about by Dalton, Avogadro and others changed the way chemists approached the problem of making chemicals in the laboratory.

To many scientists at this time, chemistry still had a mystical component – a legacy of alchemy – that was especially relevant to the so-called 'organic' compounds. Today the term organic is used by chemists to refer to the chemistry of carbon compounds (and, outside science, to food grown and raised without synthetic chemicals), but in the early nineteenth century the word organic simply meant 'derived from living matter'. Accordingly it was believed that only living things could make organic compounds, and everything of mineral origin was termed inorganic.

These organic compounds were thought to carry with them a 'vital force'; perhaps this was what Brand was seeking from his 60 buckets of urine. If compounds were simply different atoms stuck together in molecules, as Dalton and Avogadro suggested, then where could this vital force reside?

In 1828 the German chemist Friedrich Wöhler set out to eradicate the notion of a vital force by attempting to make an organic compound without the use of either living or once-living matter. Wöhler succeeded in synthesizing urea, the compound used by the body to excrete nitrogenous wastes, using only simple 'inorganic' materials.

The idea of a vital force, the mysterious essence which gives life to all living things was discredited, and life was revealed for what it was – the crowning achievement of natural chemistry, but nothing more than that.

FRIEDRICH WÖHLER
1800–82
Wöhler finally ended the idea that life depended on an intangible 'vital force' by synthesizing an organic chemical, urea, from simple inorganic materials.

Single link

CARBON CHEMISTRY

Organic chemistry is essentially the chemistry of carbon. Since Wöhler's day it has become an entire science in its own right, and a century of study has revealed a complex fabric. Carbon is among the most friendly of all atoms, and the four electrons in its outer electron shell make it exceptionally ready to combine firmly with other atoms. Indeed, its atomic structure means it links atoms together in long chains, rings or other shapes to form millions of different compounds, including the complex molecules that are the basis of life. The process of forming long chain compounds is called catenation. Only silica comes anywhere near matching this ability to form big molecules.

The molecules of aliphatic compounds have long chains of carbon atoms, arranged either in line or in branches. Cyclic compounds have rings; they include the benzene compounds which have hexagonal molecules made of six carbon atoms arranged in ring called a benzene ring. Benzene rings have a distinctive aroma, so chemicals with benzene rings are called aromatics. Many vivid 'aniline' dyes are aromatics.

It has become clear that many carbon compounds are formed with hydrogen. Those that contain carbon alone are called hydrocarbons, and they are among the most important of all organic groups since they form the basis of all our 'fossil' fuels such as oil, coal and natural gas. The key fuel hydrocarbons are alkanes, in which carbon atoms are linked by just a single bond. Methane (natural gas), ethane (industrial fuel), propane (bottled camping gas) and butane (lighter fuel) are all light alkanes.

The addition of oxygen in the form of hydroxyl groups (hydrogen-oxygen pairs) turns carbon compounds into the vast group of alcohols. These include not only ethanol or ethyl alcohol, the magic liquid that makes you inebriated, but a whole variety of others used for anything from solvents to essential oils in perfumery.

BELOW LEFT **The alkanes are one of the most important groups of carbon compounds. They are all hydrocarbons, which means they are essentially strings of carbon atoms with hydrogen atoms attached. All are found naturally, dissolved in petroleum.**

BELOW **Unlike the chains of aliphatic carbon compounds, phenyl groups are cyclic, which means their atoms are arranged in a ring. Phenyl groups are very simple parts of molecules made from just six carbon atoms and five hydrogen atoms.**

CARBON CHAINS

Methane (natural gas)

Propane (camping gas)

Butane (lighter fuel)

Ethane

CARBON RING

Phenyl group

Carbon

Hydrogen

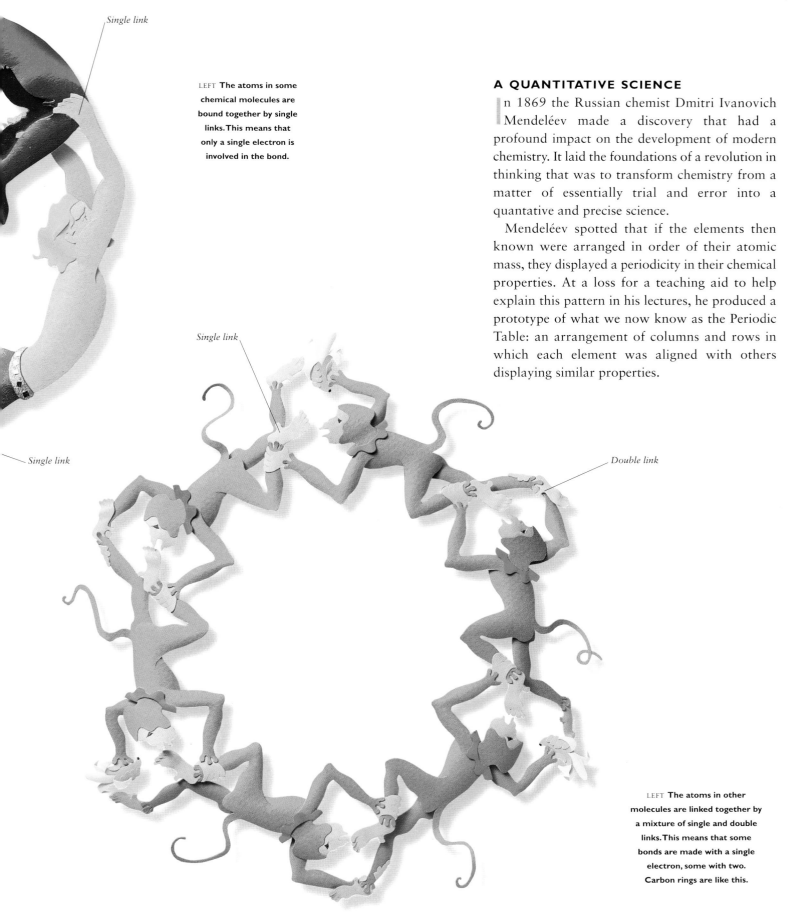

Single link

LEFT **The atoms in some chemical molecules are bound together by single links. This means that only a single electron is involved in the bond.**

Single link

Single link

Double link

A QUANTITATIVE SCIENCE

In 1869 the Russian chemist Dmitri Ivanovich Mendeléev made a discovery that had a profound impact on the development of modern chemistry. It laid the foundations of a revolution in thinking that was to transform chemistry from a matter of essentially trial and error into a quantative and precise science.

Mendeléev spotted that if the elements then known were arranged in order of their atomic mass, they displayed a periodicity in their chemical properties. At a loss for a teaching aid to help explain this pattern in his lectures, he produced a prototype of what we now know as the Periodic Table: an arrangement of columns and rows in which each element was aligned with others displaying similar properties.

LEFT **The atoms in other molecules are linked together by a mixture of single and double links. This means that some bonds are made with a single electron, some with two. Carbon rings are like this.**

wide acceptance. Anomalies in the sequence of atomic masses were also ironed out by the realization that the sequence represented the atomic number – the number of protons in the atomic nucleus – and not the atomic mass which is the number of both protons and neutrons.

Although the Periodic Table has subsequently been modified and extended, this work has only confirmed the essential truth of Mendeléev's theory. When, in 1955, American scientist Glenn Seaborg synthesized an entirely new element with an atomic number of 101, he named it mendelevium in honour of Mendeléev.

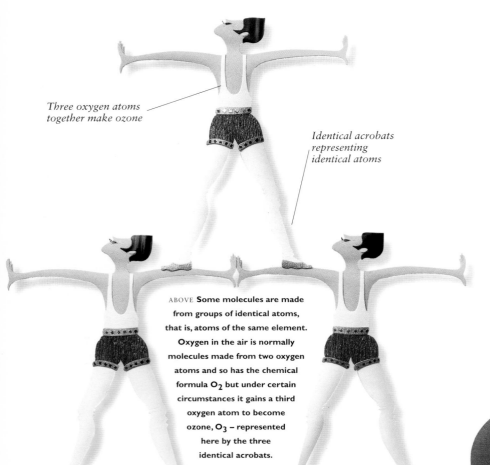

Three oxygen atoms together make ozone

Identical acrobats representing identical atoms

ABOVE **Some molecules are made from groups of identical atoms, that is, atoms of the same element. Oxygen in the air is normally molecules made from two oxygen atoms and so has the chemical formula O_2 but under certain circumstances it gains a third oxygen atom to become ozone, O_3 – represented here by the three identical acrobats.**

Hydrogen

Hydrogen

Oxygen

In compiling his table, Mendeléev encountered a problem. If the elements were simply placed in sequence in the rows, some of them did not quite fit the vertical pattern. But rather than jettison his scheme, Mendeléev placed the elements where they did fit, and left gaps between them where necessary. This was a stroke of genius, for by so doing Mendeléev was assuming that there were elements yet undiscovered that would slot neatly in the spaces. He even went so far as to predict the atomic masses of these unknown elements, and outline their chemical properties, prompting other chemists to look for 'new' elements that would plug the gaps.

Mendeléev's basic theory was proved right when scientists discovered gallium (1875), scandium (1879) and germanium (1886), which fitted in exactly where he had predicted. With this vindication Mendeléev's system began to gain

RIGHT **Molecules of compounds are made from different atoms. Water molecules, for instance, are made from one oxygen atom – represented here by our bodybuilder – and two hydrogen atoms, represented by the bodybuilder's dumbells.**

THE UNDERLYING PATTERN

Since Mendeléev's time we have pieced together exactly why the Periodic Table is such an effective tool. The basic unit of each element is the atom, and the only real difference between one element and another is the number of protons in the atom's nucleus.

Each element has a particular number of protons in the nucleus of its atoms: its atomic number. Carbon atoms have six protons, so carbon has an atomic number of six; gold has 79 protons, so its number is 79. The Periodic Table, very simply, arranges the elements in order of the number of protons in their atomic nuclei, starting with hydrogen which has just one. The more protons there are in the nucleus, of course, the heavier the atom is, in general, so atoms get bigger and heavier from the top of the table to the bottom, with very few exceptions.

Since negatively charged electrons are held to the atom by the positively charged protons, the number of protons in an atom usually determines how many electrons it has – and this in turn determines how it reacts with other atoms. Electrons stack up around the nucleus in a series of shells, but there is a maximum number in any particular shell. Once each shell is full, any extra electrons start a new shell. An atom with a full outer shell is very stable and unreactive.

Atoms with just one or two electrons in an outer shell are very reactive, because they try to either acquire electrons from the atoms of other elements to fill their quota, or get rid of their surplus by finding an atom that is short of an electron or two.

It is now clear that the Periodic Table reflects this very precisely. The columns or Groups in the table simply reflect the number of electrons in the outer shell of the atom. Every element in a particular Group has the same number of electrons in its outer atomic shells and behaves in a similar way chemically.

The rows or Periods correspond, to a certain extent, to the number of electron shells: there can be up to seven electron shells around the atomic nucleus, and there are seven Periods. As you move element by element from left to right along a Period, from Group to Group, electrons are added one by one. Each period starts on the left with a highly reactive alkali metal, with just one electron in its outer shell, and ends on the right with a stable noble gas with a full eight electrons.

EXTENDING THE TABLE

Since then, chemists have extended the Periodic Table by creating more and more synthetic elements in nuclear reactors and particle accelerators. Seaborg himself went on to make plutonium and more than nine other 'artificial' elements. The present Periodic Table has well over 100 elements, and would be virtually unrecognizable to Mendeléev. The intricacies of the modern table allow for groups within groups, which help explain the periodic properties of the most recently discovered elements – the radioactive and short-lived heavy metals. Mendeléev simply did not know to leave space for the likes of neptunium, berkelium or element 106 – referred to until recently as unilhexium but recently renamed seaborgium in honour of the scientist who laid the ground rules for creating these elements.

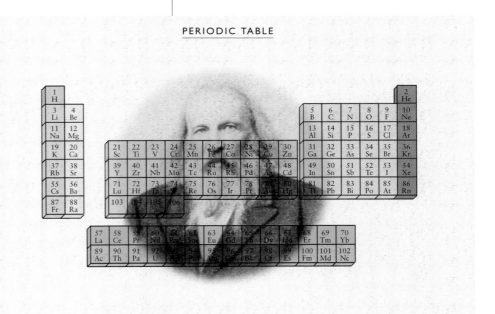

PERIODIC TABLE

ABOVE **One of the great advances in chemistry was the invention of the periodic table by Dmitri Mendeléev. The periodic table so clearly established the relationship between all the elements that it was possible to predict the existence of elements even** before they were discovered. The table lists all the elements according to the number of protons in their atom – their atomic number– in rows from left to right called periods and in columns called groups. Chemicals in each group have similar characteristics.

MOLECULAR MIRRORS

Once scientists had achieved an understanding of the nature of molecules, and how atoms fitted together to form them, they were confronted with various puzzles.

During the nineteenth century some substances were discovered to have a strange effect on a beam of light. A beam of light can be polarized by passing it through a filter that allows only waves in a single plane – say horizontal – to pass through. If the emerging beam is then shone through a solution of certain types of compound this plane of polarization is twisted through an angle. So if it starts out horizontal it might be tilted 30 degrees above or below this plane depending on the type of solution. Tartaric acid – a by-product of wine fermentation – was among the compounds that exhibited this twisting effect on light. Nobody could explain why.

Chemists thought tartaric acids were the same as compounds called racemic acids. They reacted with other substances in exactly the same ways, they tasted the same and there were no differences in their physical properties. The only logical conclusion was that they were in fact the same compound, but for the fact that tartaric acid twisted polarized light while racemic acid did not. Various scientists attempted to unravel the puzzle, but it took a scientific giant – Louis Pasteur – to take a closer look and come up with a solution.

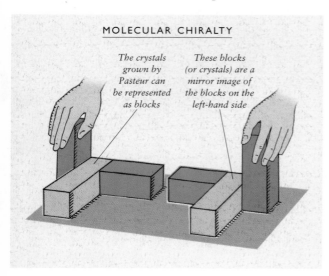

MOLECULAR CHIRALTY

The crystals grown by Pasteur can be represented as blocks

These blocks (or crystals) are a mirror image of the blocks on the left-hand side

Pasteur, well known for his medical research and the pasteurization process named after him, reacted tartaric and racemic acids in separate beakers and grew crystals of their sodium ammonium salts. When he looked at the crystals under his microscope he saw that there were crystals of two different shapes in the racemic acid, which were mirror-images of each other, and that there were equal amounts of both. There were no mirror-images among the tartaric acid crystals, though. Even more puzzling was the fact that the crystals formed by the tartaric acid were identical in shape to one set of the crystals formed by the racemic acid.

CHIRAL MOLECULES

Using tweezers and a very steady hand, Pasteur separated the different-shaped crystals and used them to make separate solutions. When he shone polarized light through the solutions he found that they twisted the light in opposite directions. Pasteur reasoned that the difference between them must be due to the way the light passed through the individual molecules, and that mirror-image crystals must be made from mirror-image molecules of the same compound. These mirror-image molecules are known as chiral molecules, from the Greek for 'hand', because some are left-handed and some are right-handed. There is no way that they can be superimposed one onto the other.

For example, the amino acid alanine, which is used by the body to make one of the B group vitamins, has complex molecules each consisting of an amino group, a methyl group, a hydrogen atom and a carboxylic acid group, all attached to a central carbon atom. If you swap any two groups – the amino and the methyl groups, say – the resulting molecule becomes the mirror-image of the original.

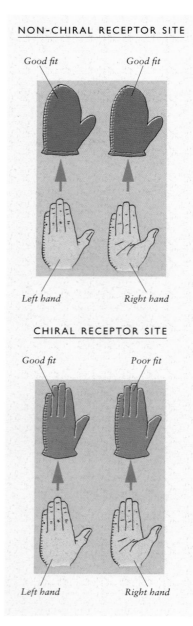

NON-CHIRAL RECEPTOR SITE

Good fit   Good fit

Left hand   Right hand

CHIRAL RECEPTOR SITE

Good fit   Poor fit

Left hand   Right hand

ABOVE The existence of mirror-image molecules has important effects in the body. Some nerve receptors, for instance, are non-chiral, that is they accept both left- and right-hand molecules. Others are chiral, which means they only accept one of the versions.

LEFT Louis Pasteur, known for discovering pasteurisation, also discovered that sodium amonium salts form two different molecules which are mirror-images of each other.

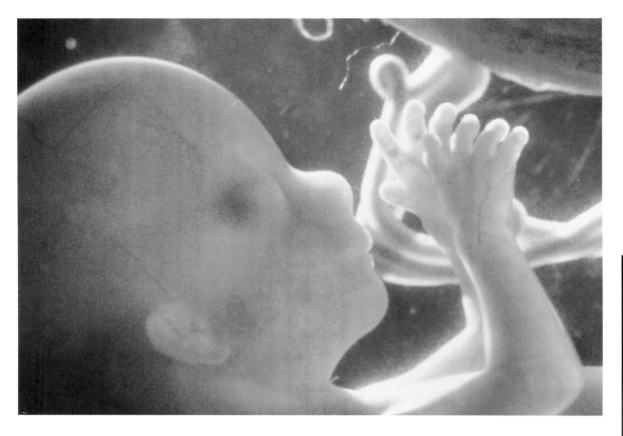

LEFT **In the human body there are many receptor sites which are chiral – and respond in different ways to left-handed and right-handed chemical molecules. Particular chiral molecules play a crucial part in the foetus as it develops normally in the womb.**

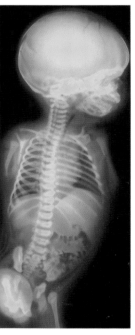

ABOVE **The drug thalidomide showed the tragic consequences of ignoring the chiral response of receptors within a developing foetus. In one-handed form the drug reduced morning sickness in pregnant women; in its other form, it had terrible effects on the unborn baby.**

MIRROR MOLECULES IN THE BODY

Intriguingly, the naturally-occurring amino acids in living things are nearly always of the same 'left-handed' form. There are many theories about the origins of this natural chirality, invoking influences such as radioactivity or even meteorite activity. Whatever its origin, however, chirality has consequences that extend way beyond a few curious optical effects. In particular, it is very important when one chiral molecule meets another in a chemical reaction.

The scent receptors that are found in your nose are chiral protein molecules. These receptors have a pocket in their surface into which smelly molecules slot like a hand in a glove. As a result, a chemical reaction is triggered that sends the scent signal to the brain. So when different-handed versions of the same molecule attempt to slot into a receptor, they trigger different reactions. One form of the chemical limonene, for example, smells of lemons, but its mirror-image molecule smells of oranges.

In medicine, the phenomenon of chirality has had serious, even tragic consequences. The drug thalidomide is one example. In one-handed form, thalidomide reduces morning sickness in pregnant women. Its mirror-image, however, has devastating effects on the development of the unborn baby.

Following the thalidomide tragedy great efforts were made to develop single-handed versions of drugs in order to prevent such a thing happening again. Thalidomide itself could not be produced in any safe form because the body converts one chiral form into the other. However not all chiral drugs are transformed in this way and so controlled chirality can also be used to increase the efficiency of a drug. The painkiller ibuprofen, for example, is a chiral substance; it has no serious side effects in either form, but one is three times more effective than the other. Ensuring that the drug has the most efficient molecular form maximizes its efficiency and reduces the dose that is needed.

LEFT, RIGHT, LEFT, RIGHT...

Because chiral molecules of different handedness often react very differently with other chiral compounds, chemists are always looking for ways to make drugs and other chemicals such as pesticides in just one form.

The simplest method is to make both chiral forms, and then separate them using a filtering apparatus that can distinguish between each form because it is chiral itself. Industrial chemists can separate compounds on the basis of how quickly they pass through a material with a chiral compound embedded in it. If a mixture of chiral forms is flushed through a column of the material one form will stick to the chiral compound more strongly than the other and so take longer to be washed through. The chemist can tap off the first chiral form as it emerges from the column, and trap the second in a separate container.

Systems like this produce the goods, but they also waste half the original material. So chemists are looking for more efficient ways of making just one form of the molecule from the start. One way is to use so-called chiral auxiliaries. These are chemical groups that come in two mirror-image forms themselves, and so add chirality to any molecule they are attached to. At Oxford University, Stephen Davies and others are perfecting ways of quickly adding chiral auxiliaries to their starting materials so they can force a reaction down one of its two possible paths to create just one handed form. They can then remove the chiral auxiliary, leaving a pure chiral product.

Other researchers are using enzymes, nature's chiral catalysts. Enzymes are proteins made of amino acids strung together and folded into a three-dimensional structure; this three-dimensionality makes them chiral. They usually act on only one handed form of their target molecule, so if the chemist's starting material is a mixture of chiral forms, treating the reaction with the appropriate enzyme will lead to just one chiral form of the product.

RINGS AND CHAINS

While some molecules are left- or right-handed others can have a variety of forms. In the latter part of the nineteenth century a German scientist – Emil Fischer – took some of the most important steps in improving our understanding of carbon chemistry by working out the structures of a variety of sugar molecules. He found that a molecule of sucrose, for instance – the sugar that many of us stir into our coffee – is built up of a glucose and fructose molecule linked together. Glucose and fructose are examples of so-called simple sugars, which can be described as rings of five carbons and an oxygen atom with various hydroxyl (hydrogen-oxygen) groups attached at different points.

Simple though that might sound, Fischer discovered that there are 16 different ways of arranging these hydroxyl groups around the sugar ring, and that each different arrangement produced a different sugar. In 1888, when Fischer was doing his research, scientists knew that glucose was such a ring, but they did not know which of the possible 16 arrangements of hydroxyl groups it had.

Fischer opened up the ring of carbons and oxygen in glucose to make an open-chain structure. He then carried out a systematic series of simple chemical reactions that would tell him where the hydroxyl groups were positioned on the open chain, and which way they were pointing.

EMIL FISCHER
1852 –1915
Fischer was the German scientist who discovered the structure and shape of simple sugar molecules. This has proved to be one of the most important discoveries in organic chemistry.

BELOW Fischer's great discovery about the molecular structure of sugars centred on the variations on a simple ring structure. Each sugar has the same basic ring of five carbon atoms and an oxygen atom. But hydroxyl groups, that is hydrogen and oxygen groups, are attached in 16 different ways to this basic ring – and each way gives a different kind of sugar.

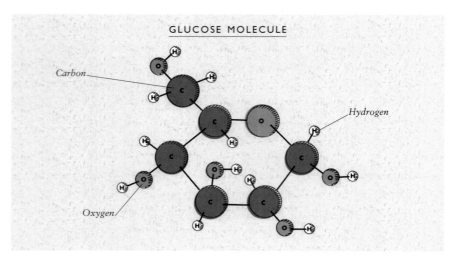

GLUCOSE MOLECULE

Carbon

Hydrogen

Oxygen

ABOVE **Our bodies are equipped with a number of different "immune" responses for dealing with invaders. These include the white blood cells, which engulf and destroy infected cells when they are alerted by the presence of antibodies, which latch onto alien protein fragments. Here a white blood cell is engulfing a ball-shaped foreign body. Once engulfed, the invader will be digested by enzymes.**

SUGAR MOLECULES

By logical extrapolation backward, he could then reconnect the ring and demonstrate the arrangement in the actual sugar molecule.

There are numerous different kinds of sugar and they play a vital part in all living systems. Both plants and animals, for instance, use the sugar glucose to provide energy. The energy is released as complex chain reactions—typically involving oxygen—which break the glucose down within the organism. Plants usually store glucose as starch; animals store it as glycogen.

Fischer's work led to an understanding of sugar structure that is becoming more and more important as discoveries are made about the effects of sugars on the body. For example, sugars form bridges between protein strands as we age, making our tissues less flexible. A similar reaction takes place when many foods are cooked—so in a way aging is a slow-cooking process. On the other hand sugars are vital to our survival.

Glucose is the sugar carried in the blood, delivering energy to the cells for maintenance and growth. Our bodies also use chains of different sugar molecules, called oligosaccharides, for many different jobs such as slowing blood cells down when tissue becomes inflamed, allowing healing to take place.

SUGAR STRINGS

Among their many other functions in the body, sugars help coordinate our immune systems to destroy alien tissue, if it should get inside. According to Oxford University sugar expert Raymond Dwek, 1 percent of the antibodies floating in the bloodstream are targeted at one particular sugar found in the tissues of all animals except ourselves. This sugar is the key to successful xenotransplantation—the transplantation of an organ from another species into a human. Its presence in animal tissue means that when an animal organ is transplanted into a human, the 1 percent of our antibodies targeted to the sugar are almost instantaneously activated to seek out and destroy the alien tissue. Assuming there are no ethical objections to the transplant, finding a way to block this fatal reaction helps a patient's body accept the donor tissue and save his life.

In studying these sugars, though, scientists run into a major problem—they are very difficult to make in the laboratory and therefore extremely expensive. Rather than a single ring like glucose—a monosaccharide—the antibody-activating sugars are made of several different rings in a chain—an oligosaccharide. This means that the sixteen possible arrangements of hydroxyl groups in glucose can be multiplied by three, four, five, and so on. According to one expert—David Crout of Warwick University in England—each additional sugar unit in a chain can take several weeks of work, and one particular oligosaccharide costs £5 million per gram. This is one of the most expensive commodities around.

Crout and his team have devised a method of cutting the costs using nature's enzyme-based catalysts to speed up and control the synthetic process. The basic problem is that mixing sugars in a test tube results in a syrupy mess rather than a sweet and pure molecule. Nature, of course, has worked out a solution to the problem in the form of enzymes. Enzymes control how successive sugar molecules are added to a growing chain so that only the required sugar chain is formed in an animal's cells.

ENZYMES

The Warwick team decided that, rather than spend their lives trying to find ways to build sugar strings chemically, they would follow nature's example and use enzymes, but in reverse. They have been studying a group of enzymes known as glycosidases. In living organisms these normally break sugar strings apart, for recycling and disposal by the cell as part of everyday metabolism, but the researchers found that under the right chemical conditions they could make the enzyme force two sugars to join together instead. Their work could significantly reduce the cost of synthesizing these complex sugars, with profound – if controversial – implications for the future of tissue transplants.

CATALYSTS AND ENZYMES

Catalysts are the introduction agencies of the chemical world, easing the relations between elements and compounds and encouraging reactions – but without actually getting involved themselves. Without them the vast majority of manufactured compounds could not be made.

Every reaction needs a certain amount of energy to get it going, called the activation energy. A candle won't burn, for instance, until you light it with a match. Catalysts work by providing an alternative route for a reaction and lowering the activation energy. In a rocket, for instance, the combination of hydrogen and oxygen that provides the thrust depends on a metal catalyst, typically platinum.

The living world depends totally on natural catalysts called enzymes. They are large, complex proteins, and each is highly specific, working for only a particular kind of reaction. They are denatured – damaged – in temperatures above 60°C, and in excessively acid or alkaline conditions. Biological washing powders, which rely on enzymes, use a less than hot wash not just to save energy, but because excessive heat destroys their enzymes.

ABOVE **A lighted match provides the energy needed to get going the chemical reaction of a candle burning.**

ANALYSIS AND SYNTHESIS

Chemists now have a multitude of techniques at their disposal for manipulating carbon compounds, from simple methane to complex sugars and polypeptides (the basis of proteins). But it was not always so easy.

The real breakthrough in organic, or carbon chemistry came in the 1960s, when Harvard University professor and Nobel Laureate Elias J. Corey developed a methodology called 'retrosynthetic analysis'. The technique has become one of the most powerful tools at the chemist's disposal for building large complex organic molecules from smaller, more readily available – and usually considerably cheaper – starting materials.

The method works by picturing a target molecule as if it were a jigsaw puzzle. By working back from the target it is possible to look for chemical components that, when reacted together, will assemble into the complete puzzle under the direction of catalysts and reagents.

One such chemical puzzle has received world-wide attention. Taxol is a natural compound extracted from the bark of the Pacific yew tree Taxus brevifolia. It acts as a natural defensive weapon for the tree, arming it against pests and diseases, but it is also an effective defence against some forms of cancer in humans. Test results were so encouraging that scientists began to demand samples of the chemical so they could study how it worked.

There was a problem, however. Even a hundred-year-old yew makes only tiny amounts of Taxol – about 300 milligrams from three kilograms of bark – and this is not nearly enough to meet the growing demand. As a result Taxol is extremely expensive, so while some researchers looked for alternative sources such as the needles of the European yew, chemists began looking for a way of making Taxol in the laboratory.

The compound is basically made up of carbon ring systems with several different chemical groups dangling around each ring. By using retrosynthetic analysis to work out which component parts

POLYMERS

Polymers are substances made from thousands of small carbon-based molecules called monomers which are strung together in long chains. Most plastics are polymers made from ethene, one of the products of oil. When heated under pressure, the ethene molecules join into chains with 30,000 or more links. These long molecules get tangled like spaghetti, and it is the way these strands are tangled that gives the plastic its strength or pliability. If the strands are held rigidly together, the plastic is stiff, like Formica. If strands can slip easily over each other, the plastic is flexible like polythene. And forcing the molecules through tiny holes lines them up to create fibres such as nylon.

Plastics are man-made, but some polymers occur naturally. Sugar polymers – which are carbohydrate polymers made from chains of carbon, hydrogen and oxygen – play a key role in energy exchanges in both plants and animals. Plants store energy as the sugar polymer starch, a complex 'polysaccharide' made from thousands of carbohydrate units. Animals store it as the sugar polymer glycogen. These polymers can be broken down by enzymes into simpler sugars such as glucose. They can also be linked together in different ways to form structural polymers. Cellulose is one such natural polymer, from which all plant cells are made. It is the basis of wood and cotton fibre, so both are natural plastics.

ABOVE **When eating an ice cream our bodies use chemical enzymes to break sugar strings apart. Chemists use them to build strings up.**

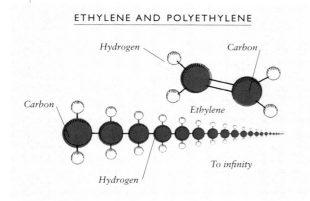

ETHYLENE AND POLYETHYLENE

Hydrogen
Carbon
Carbon
Ethylene
Hydrogen
To infinity

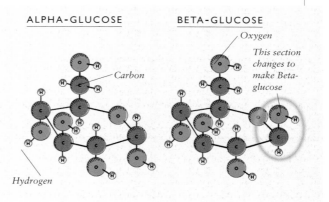

ALPHA-GLUCOSE

Carbon
Hydrogen

BETA-GLUCOSE

Oxygen
This section changes to make Beta-glucose

might fit together to complete the picture, two research groups – one led by Kyriacos Nicolaou at the Scripps Research Institute in California, and one led by Robert Holton of Florida State University – have finally come up with several different ways of building Taxol from much simpler chemical components.

The two groups used slightly different approaches. Holton's group started with a cheap precursor compound, camphor, and built up the complex structure through many reactions in sequence. Once they had the ring system they added the side chains. Nicolaou's team, on the other hand, used a convergent approach. They made the two main rings, added the various chemical groups and then coupled them together. Both approaches sound rather simple when described in this way, but the many reactions involved took years to perfect.

Such research has a valuable extra dimension, for having sorted out Taxol itself, chemists can make variations on the main theme that might actually be more convenient and effective, with fewer side effects. They might also be a great deal cheaper, bringing much needed relief to cancer sufferers worldwide.

ABOVE FAR LEFT **Polymers such as ethylene and polyethylenes are molecules formed by long chains of carbon atoms, typically with hydrogen side branches.**

ABOVE NEAR LEFT **Glucose molecules, made from carbohydrate rings, may also be strung together in long chains to form natural polymers such as starch.**

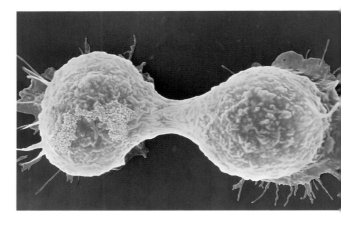

MOLECULAR VACCINES

Research groups around the world are investigating the possibility of vaccinating against diseases for which there are currently no long-term effective drugs. The new approach involves using chemicals to stimulate the body's immune system, instead of conventional vaccines. One of the nastiest diseases around – meningococcal meningitis – could one day be dealt with in this way, if current research proves successful. The problem with a 'conventional' type vaccine for this disease is that the surface of the meningitis bacterium is complex, some components are toxic and others can block the activity of protective antibodies. This means that creating a vaccine based on the bacterium, which is the usual approach, is not possible.

Molecular microbiologist John Heckels and his colleague Myron Christodoulides at Southampton University in England are trying a different strategy. They are working on proteins called antigenic peptides, which in minute amounts act as signals to the body's immune system to attack the pathogens that cause the disease. Such chemicals are known as molecular vaccines.

Molecular vaccines will not be limited to meningitis. They offer the hope of triggering our natural defences against virus-infected cells and cancer cells, and may be able to block the destructive action of cells that destroy tissue in diseases such as arthritis and multiple sclerosis. And since the body's own immune system is doing the work, there should be minimal side effects. There is a major obstacle in the way of the perfect vaccine, though: identifying the active components, which exist in such vanishingly small amounts in the first place.

Every cell in our bodies contains about 10,000 different proteins. These proteins are continually purged from the cell contents by enzymes that break them into smaller fragments, and then carry them to the cell surface where they are removed by other cells. When a cell changes, as happens in cancer or when it is infected by a virus, this clean-up process still continues, but there are new proteins being broken down within the cell – those generated by the virus or caused by the cancer.

Our immune systems have two main seek-and-destroy systems for spotting these rogue proteins on cell surfaces, and so overcoming the disease. The first involves antibodies that latch on to the alien protein fragments and send signals to white blood cells to destroy the obviously unhealthy cell. The second is based on cell-killing T cells, which keep an eye on the surface of cells and quickly spot when 'alien' protein fragments appear. The T cells then launch an attack to destroy the diseased cells.

When the system breaks down the presence of the unhealthy cells gives rise to the symptoms of the particular disease. Yet only very tiny amounts of the antigenic peptides are found on the cell surfaces, so highly sophisticated analytical chemistry techniques must be used to isolate and identify them. Luckily once they are found, the peptides are actually relatively easy to synthesize, purify and standardize, so molecular vaccines may soon become an everyday reality.

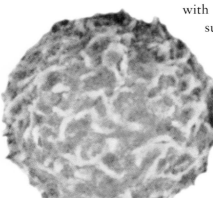

RIGHT This scanning electron micrograph shows a breast cancer cell dividing. Researchers believe that such cancers are just one of the diseases that it may be possible to develop molecular vaccines for. The vaccine will activate the body's natural defences to fight against the cancer cells.

ABOVE White blood cells are one of the key elements in the body's defence against disease. But they only go into action when triggered by antibodies on the disease-causing cells. Molecular vaccines will help the body develop the necessary antibodies to provide effective protection.

RIGHT Once antibodies alert white blood cells to the presence of invaders, they rapidly cluster around the site, engulfing the offending cells and destroying them with enzymes.

TANDEMS, CASCADES AND DOMINOES

Ideally chemists choose cheap and readily available pieces for their molecular jigsaw puzzles, which fit together leaving very small amounts of by-product. Achieving this ideal takes skill, time and energy, though, for each piece of a molecular puzzle normally needs numerous modifications to make it fit properly. This increases the potential for side reactions, leading to reduced amounts of product that then has to be purified, using more energy. Often the jigsaw-puzzle pieces are mini-puzzles in themselves that have to be sorted out. So despite the best efforts of hard-working chemists, syntheses of complex molecules often boil down to wasted materials and dirty solvents.

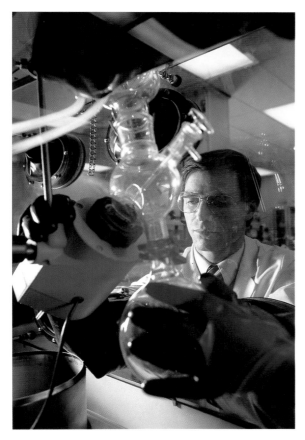

ABOVE **New chemicals can be created and tested theoretically on computer, but there still comes a time when the virtual ideas have to be tried out with real chemicals. Here a chemist is using a glove box to work on drugs based on the precious metal platinum. Such drugs include the anti-cancer drug cisplatin.**

CHEMICAL REACTIONS

A chemical reaction is a process that changes one substance into another. Crucially, it involves a coming together of different chemical elements and compounds which rearranges their constituents to form new compounds and separate different elements. On a basic level it involves the meeting of atoms, ions and molecules and the exchange of electrons between them. Although the new substances formed after a reaction must contain exactly the same atoms, very few reactions are genuinely reversible. Once coal is burnt to ash it is impossible to reform it to coal. Reactions between chemicals are essential to the function of the world, whether through the photosynthesis of plants, which turns water and carbon dioxide into sugar, or the breaking of those sugar molecules in reaction with oxygen to give the body muscle power.

Some atoms are considerably more reactive than others. Essentially, the reactivity of an element depends on its readiness to gain or lose the electrons used for bonding. Different elements can be placed in the reactivity series according to their willingness to react. Metals like potassium and sodium are high in the reactivity series and rarely exist unchanged for long. Gold is low in the series and can survive in pure form almost indefinitely.

One of the most effective ways of looking at the reactivity series is in terms of electronegativity, which is the ease with which an atom can attract electrons to itself. This depends on the atom's valency and was developed into a complete scale, predicting bond strength, by Linus Pauling in one of the chemical breakthroughs of the century. The key factor is the number of electrons in the outermost shell of the atom.

The noble gases such as argon and crypton are all unreactive because their outer shell is full. Oxygen, with four electrons missing in its outer shell, is highly reactive. Atoms with fewer electrons missing are less reactive, but still a great deal more liable to react than the unresponsive noble gases.

PHOTOSYNTHESIS

Water

Carbon dioxide

Oxygen

Sugar

ABOVE **Photosynthesis in leaves uses the Sun's energy to convert water and carbon dioxide to sugar. Oxygen is also created and released.**

RIGHT **Heterocycles are an important group of chemical molecules found everywhere in nature, occurring in thousands of compounds in plants and animals. Like all natural chemicals they are predominantly carbon-based, but their ring includes an atom other than carbon. They are often the biologically active ingredient in many household chemicals.**

CASCADES

In an attempt to streamline the process, chemists have found ways to carry out the key steps of a synthesis in a single reaction pot. The method uses a cascade of reactions, avoiding the need for separate steps with their associated waste. Such cascades have been investigated by Lutz Tietze's team at the Georg-August University in Göttingen. In one such cascade, or domino reaction, they begin with two simple carbon compounds and use a small amount of catalyst to push them into a reaction. This leads to an unstable intermediate compound, which then cascades through a reaction that closes it up to form a ring.

This is an example of a very short cascade, with only two dominoes tumbling, but it shows the vital features needed to make complex molecules cheaply. It all happens in a single fluid sequence, starting with simple materials and producing a complex molecule. What's more, the starting materials can have different chemical groups added without upsetting the cascade.

Andrew Holmes and his team at the University of Cambridge are using short cascades to make a group of molecules known as heterocycles, which have an atom other than carbon in a ring. In nature heterocycles are used for thousands of compounds in plants and animals, and they are the common cores used to make many biologically active molecules for use in pharmaceuticals and other products.

BELOW **To synthesize complex molecules in a single vessel chemists such as Lutz Tietze have developed the idea of cascades, in which catalysts provoke a successive series of chemical reactions. These chemical reactions can be thought of in terms of a domino effect.**

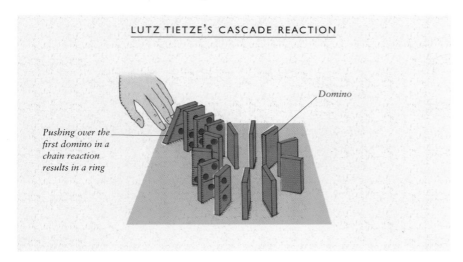

LUTZ TIETZE'S CASCADE REACTION

Domino

Pushing over the first domino in a chain reaction results in a ring

COVALENT COMPOUNDS AND METALS

The sharing of electrons in covalent compounds creates strong bonds between the atoms in each molecule—but it also means that they tend to turn a cold shoulder to other molecules. So covalent compounds are usually liquids or gases at normal temperatures, and their melting and boiling points are low because the bonds between their molecules are very easily broken.

This contrasts with metallic bonds, in which valency electrons move freely between atoms, helping to bind the huge numbers of ions together. This is what makes metals generally so solid and tough, and accounts for their high melting and boiling points.

To some extent the same is true of ionic compounds such as common salt, in which positive ions and negative ions come together to form a giant "ionic" lattice that gives ionic compounds characteristically high melting and boiling points.

Using one particular method, Holmes and his colleagues begin with a straight chain or linear molecule that has a side arm containing a nitrogen atom—this will be the heteroatom in the ring. At one end of the molecule they add a reactive double bond and at the other a triple bond. Then they set the cascade going by heating the reaction mixture. The triple bond on one end of the molecule whips around and latches onto the nitrogen atom in the side arm, forming a small ring-shaped intermediate molecule. This makes the double-bond end whip around as well, and grab the nitrogen atom to form a second ring. By adjusting the compounds with which they begin, Holmes and his team have made several new compounds with two, three, and four heterocyclic rings joined together in this way.

Cascades have the added advantage of often forming only one chiral—or handed—form of the final molecule. Chirality has become increasingly important to chemists, and the specificity of a cascade for one chiral form helps the chemist avoid additional separation steps at the end of the reaction, improving reaction yields, reducing waste, and saving energy.

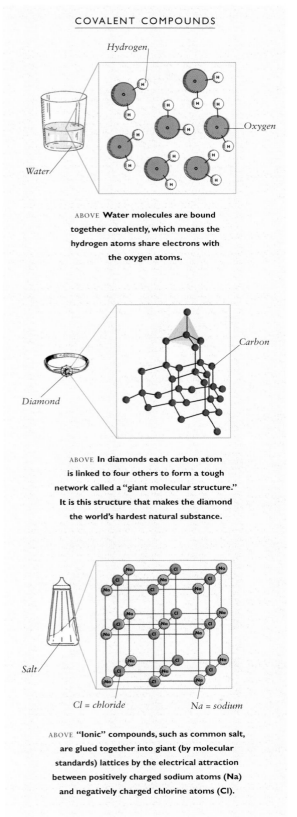

COVALENT COMPOUNDS

Hydrogen

Water

Oxygen

ABOVE **Water molecules are bound together covalently, which means the hydrogen atoms share electrons with the oxygen atoms.**

Diamond

Carbon

ABOVE **In diamonds each carbon atom is linked to four others to form a tough network called a "giant molecular structure." It is this structure that makes the diamond the world's hardest natural substance.**

Salt

Cl = chloride   *Na = sodium*

ABOVE **"Ionic" compounds, such as common salt, are glued together into giant (by molecular standards) lattices by the electrical attraction between positively charged sodium atoms (Na) and negatively charged chlorine atoms (Cl).**

BONDING EXPERIENCES

Some atoms exist in isolation for long periods, but most combine with others to form molecules. But there is a pattern to these chemical matings. Each atom combines only with a certain number of other atoms. A hydrogen atom combines with one chlorine atom but never two, while an oxygen atom can join with two hydrogen atoms. A nitrogen atom can join up with three hydrogen atoms, and a carbon atom can join up with four. In effect, each atom has a certain number of "hooks" on which it can hang other atoms, and this is expressed in the atom's valency.

Although some atoms show different valencies in different combinations, the number of hooks directly reflects the number of electrons in the atom's crucial outer shell, sometimes known as its valency shell—and this gives a clue to the way bonding actually takes place. When substances react together, the tendency is always for their atoms to gain, lose, or share electrons until they acquire a stable—that is, full—outer shell. And as they do so, they develop a mutual attraction that makes them bond.

When these bonds are made, electrons in the valency shell (valency electrons) are either lost or gained, as they are in "ionic" and "metallic" bonds, or shared with other atoms, as in "covalent" bonds. Ionic bonds form when atoms with just a few electrons in their outer shells donate them to atoms with just a few missing from theirs. The atoms losing their negatively charged electrons become positive ions, while the atoms gaining them become negative ions—so they are drawn together by mutual electrical attraction, positive to negative.

Covalent bonds occur when two atoms actually share pairs of electrons, as between two hydrogen atoms and one oxygen in water. In metallic bonds, atoms release electrons *en masse,* and the negatively charged electrons pass freely among them, binding huge numbers of positively charged ions in a giant metallic lattice.

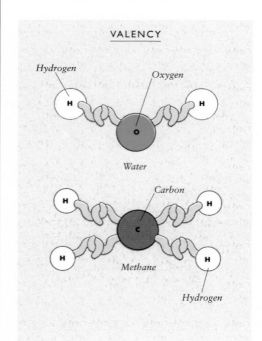

VALENCY

Hydrogen H
Oxygen H
O
Water

Carbon H
H
C
H
H
Methane
Hydrogen

ABOVE In water molecules the single hooks of two hydrogen atoms link with the double hooks of an oxygen atom. In methane, on the other hand, the single hooks of four hydrogen atoms link with the four hooks of a carbon atom.

BUCKYBALLS

In 1996, one area of entirely fundamental chemical science was awarded the Nobel Prize for chemistry, highlighting the importance of pursuing scientific research without necessarily having a technology to aim for. The prize was awarded jointly to Professor Sir Harold W. Kroto, of the University of Sussex, England, and Professors Robert F. Curl Jr. and Richard E. Smalley of Rice University in Houston, for their discovery of the all-carbon soccer-ball molecule known as buckminsterfullerene.

Carbon had until then been considered pretty much a closed book. There were two main forms, or allotropes: diamond, the stuff of rings, and graphite—the "lead" in a pencil. There were also several noncrystalline, irregular forms known as amorphous carbons. But in 1985 Kroto, Curl, and Smalley discovered that carbon could exist in an entirely new allotropic form, made up of tiny spheres of carbon, and that these spheres represented an entirely novel molecular form.

The team had used an analytical technique known as mass spectrometry to look at certain samples they had produced, and the results of the analysis hinted at the presence of heavy cluster molecules with very high symmetry. After many puzzling hours, the team eventually figured out that the high symmetry could only be explained if the main molecule was a sphere of sixty carbon atoms. It turned out that the actual shape is a mathematical solid known as a truncated icosahedron—a sphere with twenty hexagonal faces and twelve pentagonal faces. This is exactly how a modern soccer ball is constructed. They named the molecule buckminsterfullerene, after the architect Buckminster Fuller, who employed the structure in his buildings, but the soccer ball analogy has proved irresistible and the molecules are generally known as buckyballs. Buckyballs are one of the most exciting discoveries in recent chemical history. Although they have found very few uses as yet, most chemists are convinced that they will prove to be the building blocks of many important chemicals in the future.

*The balls represent
carbon atoms*

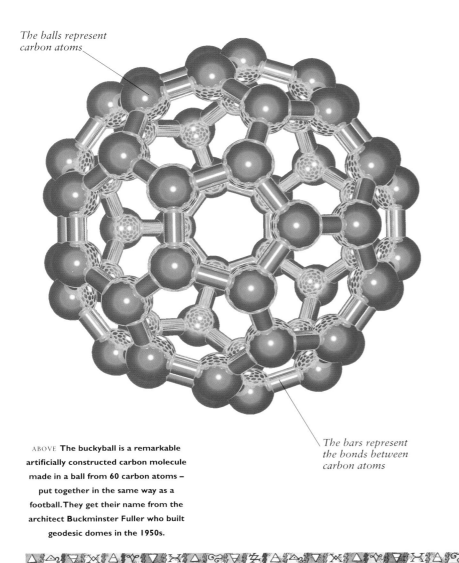

*The bars represent
the bonds between
carbon atoms*

ABOVE **The buckyball is a remarkable
artificially constructed carbon molecule
made in a ball from 60 carbon atoms –
put together in the same way as a
football. They get their name from the
architect Buckminster Fuller who built
geodesic domes in the 1950s.**

BOND ENERGY

Nearly all chemical reactions involve a change in energy, as bonds are
broken and forged. For a bond to break, energy is needed, while making
a bond releases energy. The greater the strength of the bond, the greater its
bond energy – that is, the more energy is needed to break it. Some reactions
involve electrical energy, some involve light, but most involve heat.

A reaction which releases heat to its surroundings is described as
exothermic. When wood burns, for instance, more energy is released by the
making of bonds than is absorbed to break the bonds, so the reaction gives
off heat. An explosion is the most intense form of exothermic reaction. A
reaction which absorbs heat from its surroundings is described as
endothermic. This is how the cold packs used by athletes to cool injuries
work. Inside the pack ammonium nitrate dissolves in water, a process which
involves breaking bonds. Heat energy is absorbed to break the bonds, this
makes the water cool.

THE FULLERENES

Many chemists were not convinced of the
validity of the initial results, and it was to be
another five years before scientists made the 'new'
form of carbon in the laboratory. Ironically it was
two physicists, Donald Huffman and Wolfgang
Krätschmer, who found a way of making
buckyballs in large quantities so that scientists
could kick them around the experimental arena.
They condensed carbon vapour in an atmosphere
of an inert gas. The excited carbon atoms
combined and curled up into clusters with up to
several hundred atoms, and some of the clusters
were buckyballs.

Chemists have now managed to carry out many
of their stock reactions on buckminsterfullerene
and its rugby-ball shaped cousin, [70]fullerene.
They have attached almost every known chemical
group to fullerenes, and they have also discovered
ways of trapping metals and other elements inside
the carbon spheres as they form. The effects on
the electrical properties of the materials are
astounding, making them superconductive at a
balmy 77 kelvin (just 223°K below the freezing
point of water). They have even trapped fullerenes
themselves in the hollow spaces of porous
materials such as the natural zeolite minerals, and
these composites can be made to glow with an
eerie white light.

The biology of the fullerenes is also beginning
to unfold, and researchers have made several
versions of the buckyballs that have biological
activity. For instance, laboratory studies suggest
that one derivative can block the activity of certain
enzymes of the AIDS virus HIV.

Numerous research groups are looking at ways
of building the all-carbon balls from scratch using
traditional organic synthetic techniques. In this
way they may be able to add chemical groups with
useful functions as they build the balls, rather than
trying to add them later. They hope this will give
them more control over the final product, as well
as allowing many more groups to be incorporated
into the sphere. Buckyball chemistry is still very
much in its infancy.

RIGHT **Virtual reality systems allow chemists to pick up molecules, turn them round, add or remove groups and see how they react together – all on computer. Here the large purple ribbon-like molecule is a model of an HIV enzyme, while the blue is DNA.**

VIRTUAL CHEMISTRY

Chemists have toiled for decades over their test tubes and reaction flasks searching for ways to mix and match atoms into new molecules. But the toil may soon be over, for less flask-happy chemists have designed computer programs that can learn how to design reactions and ultimately program a robot arm to shake the reaction flask for them.

Chemists are finally beginning to explore the power of the computer to work in abstract cyberspace. Instead of putting chemicals together in test tubes, they are beginning to work in a virtual world, putting computer models of molecules together to see how they react. Work is just beginning on the programmes needed to do this effectively.

Johann Gasteiger of the Institute for Organic Chemistry at the University of Erlangen-Nürnberg in Germany is one scientist who feels that chemists work too hard. He and his colleagues have spent the last 15 years designing a program which, they hope, will be able to look at a molecule and work out exactly how it will react without even raising a test tube.

The software uses the information found in databases containing hundreds of thousands of chemical reactions – each with its own reaction conditions, catalysts and reagents listed together with the physical parameters of the molecules involved. A chemist could search the database by hand or use a search program to pick out reactions of interest, but a single search can lead to a list of several hundred reactions from a database that can contain millions. Manual analysis is both laborious and time consuming.

Gasteiger felt that he could train a computer program called a Kohonen neural network to do

the sorting for him. Gasteiger reasoned that if all the physico-chemical properties of reactions could be fed into the program from a database they could create a complete map of chemical reactions, analogous to the network of sensory inputs inside our brains. Instead of sensory inputs, Gasteiger's team homed in on all the factors affecting a chemical reaction, such as temperature and pressure.

Gasteiger's team picked on a single broad class of chemical reactions to test their ideas, and used a search program to look for a general type of reaction. They obtained a set of 120 reactions from a database of 370,000. To simplify things further, they chose seven characteristic physical properties associated with the reaction centre of each molecule as the input for the neural network.

CHEMICAL REACTIONS

When a reaction is input the variables are mapped onto a 12 by 12 grid of artificial neurones in the software. The reactions are input sequentially, and after each entry the weights on each neurone are adjusted to make them more like the input variables. The closer the hit on each neurone, the higher the adjustment. This 'trains' the network. If the next input reaction has similar variables it is mapped onto a neurone close to the first, but if the value is different, it is placed on a distant neurone and the weights have to be adjusted again.

The process creates a two-dimensional 'landscape' of reactions, in which similar reactions are clustered together while dissimilar reactions lie far far apart from each other. Isolated peaks in the landscape point to unusual and uncommon chemical reactions.

CAMEO AND SYNGEN

The most exciting aspect of this is the way it can predict the properties of new compounds. When the seven variables for a new molecule are fed into the trained neural network, it processes them and assigns a particular region to the new molecule. This allows the chemist to see – within half a second – the likely reactions a molecule will undergo in the lab.

William Jorgensen of Yale University in New Haven Connecticut is working on another program called CAMEO (Computer Aided Mechanistic Evaluation of Organic reactions). The chemist feeds the starting materials into the computer and adds the reaction conditions using drop-down menus. The program then attempts to predict what will happen, using a set of rules based on decades of laboratory observations to string together the fundamental steps to create known and perhaps unknown reactions. For instance, Baldwin's rules, invented by an Oxford University chemist, predict whether or not a carbon chain with an oxygen or nitrogen along its length will curl up to form a ring or not. Cram's rules – devised by American chemist Donald Cram – can predict which way a reagent will attack a carbon-oxygen double bond.

There are many routes possible to a molecule of interest. James Hendrickson of Brandeis University has devised yet another program called SYNGEN – for synthesis generator – which looks for the shortest path by dismantling the reaction target on the computer screen. For each skeletal dissection, it tells the chemist which functional groups would be necessary to initiate a reaction sequence to rebuild the skeleton, and make the product.

Used together, computer programs may enable chemists not only to classify reaction types and predict how a particular molecule may react, but also to work out what type of reaction may be needed to produce a newly designed molecule. The neural network might classify the reaction, CAMEO could find out whether other reactions might work, and SYNGEN could optimize the route and figure out the budget.

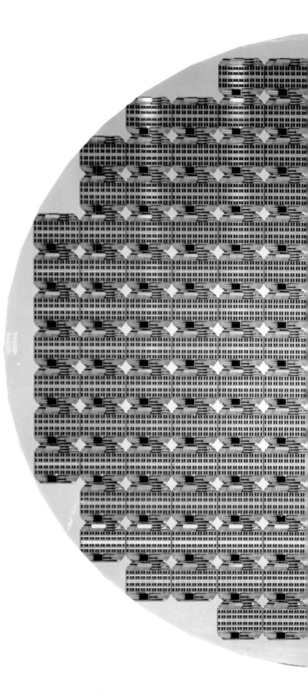

BELOW **The computer microchip is the test tube of the future – the place where chemists try out chemical reactions.**

A CHEMICAL COMPUTER

Chemists are not content with simplicity. Despite the advances in computer-aided molecular design many chemists are now moving away from single molecules and into the field of molecular architecture. This involves joining individual molecules to build sub-microscopic molecular machines or 'supermolecules': the building blocks of computers based on chemicals.

The idea of building a computer from components on the molecular scale, rather than relatively enormous chunks of semiconductor materials such as silicon, appeals greatly to modern chemists. It provides them with the opportunity to create some beautiful chemical structures, solve extremely complex synthetic puzzles and lead the way to a future of super-computers the size of a paperback book.

An electronics designer tends to work from the top down, to make devices ever smaller. This approach has given us the 'silicon chip' microprocessor: a wafer-thin section of synthetic crystalline semiconductor material with micro-scopic components – transistors, diodes and capacitors – etched onto the surface using printing techniques and corrosive chemicals.

The chemist, on the other hand, has the tools and techniques to work from the bottom up, first designing the device and then assembling it from its component parts. The tools and techniques a chemist can apply are the basis of a rapidly growing branch of synthetic chemistry called supramolecular chemistry – a field that could bring chemistry to the electronics industry in a big way.

SUPRAMOLECULAR CHEMISTRY

Jean-Marie Lehn of the University of Strasbourg, France, shared the 1987 Nobel Prize for chemistry with Americans Donald Cram and Charles Pedersen for their pioneering work with designed molecular complexes. Lehn defines supramolecular chemistry as, 'Chemistry beyond the molecule ... the designed chemistry of intermolecular bonds'. What he means by this is that, whereas normal everyday chemical reactions involve the breaking and making of bonds between atoms, supramolecular chemistry is based on the making and breaking of bonds between whole molecules.

These so-called non-covalent bonds – as opposed to the common covalent bonds in organic compounds – are at the heart of many of the computational-type reactions carried out millions of times a day by living organisms. With planning and design, the elaborate structures that supramolecular chemistry makes possible could hold the key to building devices on the nanoscale (billionths of a metre across) and could eventually lead to the creation of a computing machine based on chemistry rather than electronics.

There are two major principles involved in supramolecular chemistry. Both borrow heavily from nature for their inspiration, but are, on the whole, working with synthetic systems.

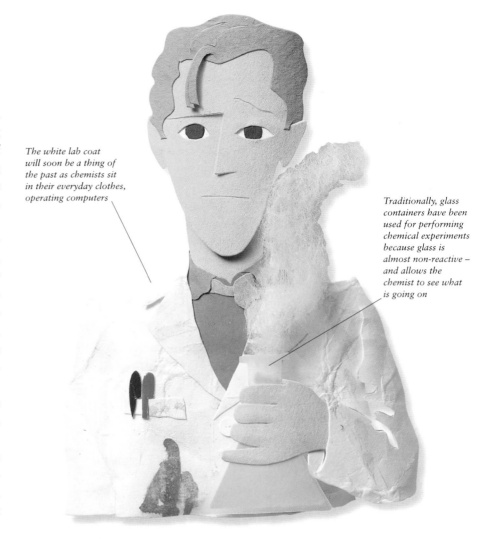

The white lab coat will soon be a thing of the past as chemists sit in their everyday clothes, operating computers

Traditionally, glass containers have been used for performing chemical experiments because glass is almost non-reactive – and allows the chemist to see what is going on

ABOVE **There is still a place for the traditional chemistry laboratory and many chemists still prefer 'hands-on' chemistry involving real reactions and materials.**

RECOGNITION AND SELF-ASSEMBLY

The first of these principles is the property of some compounds to undergo molecular recognition. This process is analogous to many biochemical events, such as a molecule binding to a receptor or slotting into an enzyme's active site like a key in a lock. The supramolecular chemist builds rings and cages, for instance, that can recognize and trap smaller molecules or ions in their cavities just as a scent molecule fits into a scent receptor. In the 1960s and 1970s Charles Pedersen designed a group of compounds known as crown ethers which were among the first synthetic chemical systems known to recognize other chemical species and so bind to them selectively. One of the smallest crown ethers, for example, recognizes and traps ions of the alkali metal lithium in preference to other alkali metal ions in solution.

BELOW **Modern chemistry now typically involves assembling atoms and simple molecules to create new molecules in a virtual world on computer. Eventually even computers themselves may be created by assembling molecules designed on computer.**

On computer new molecules can be tried and tested safely, in an instant

Sophisticated computer programmes allow the chemist to turn the molecule and view it from all angles

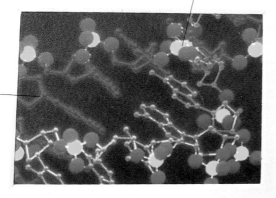

ALFRED NOBEL

Alfred Nobel is chiefly remembered for funding the Nobel prizes which bear his name. The prizes were intended to honour people who helped humankind, but, ironically, the money to fund them came from the manufacture of explosives and other armaments. In fact, Nobel's preoccupation with explosives led to personal tragedy in 1864 when his factory began to make nitroglycerin and his brother Emil was killed in an explosion. Rather than abandon explosives, Nobel invented a safer substance, dynamite. Ten years later, in 1875, he invented gelignite, still one of the most widely used explosives. Nobel had hoped that the destruction he manufactured would put an end to war in the same way that people a century later were seduced by the nuclear deterrent. His funding of the Nobel prizes was an attempt to rectify his mistake.

The second principle of supramolecular chemistry is known as self-assembly. Again, this idea has a natural analogue in the piecing together of a virus that attacks tobacco plants. The virus is a tiny cylinder of protein with its genetic RNA held inside. The cylinder is made up of 2,130 identical wedge-shaped protein subunits, which are stacked together in a spiral form like bricks around a turret. If the virus and its long strand of RNA (which has 6,400 basic components) are broken apart chemically into individual pieces and then thrown back into a solution, they recognize each other and self-assemble in the laboratory. The result is, quite eerily, a fully infectious tobacco mosaic virus once again.

Self-assembly is inextricably linked to the idea of molecular recognition: the recognition of the subunits by each other guides the construction of the supermolecule. The crown ether known as

12-crown-4 (a 12-membered macrocycle with four equally spaced oxygen atoms) is a good example. Not only does it selectively recognize lithium ions, but simply mixing the ring's component parts in a solution containing lithium ions results in the spontaneous self-assembly of the ring around the metal ion – which acts as a kind of template for the structure.

When chemists began to apply these two principles – recognition and self-assembly – to the design of more complex chemical structures, it became apparent that they were drafting the blueprints for devices that could form the basis of molecular computers.

SHOPPING LIST FOR A MOLECULAR COMPUTER

At this stage in the development of our hypothetical molecular computer, nobody really knows what devices and components will be needed. It might be that all the electronics technology of the last half of the twentieth century would be rendered irrelevant by a new approach. According to nanotechnology pioneer K. Eric Drexler, attempting to emulate microelectronics on the nanoscale is probably not the best approach to the problem. He has suggested that the 'difference engine' of the Victorian mathematician Charles Babbage could be a more likely prototype for a molecular computer, with molecular bearings and cranks carrying out compu- tations on a scale billions of times smaller than Babbage could have envisaged.

Regardless of the machinations of Drexler, chemical scientists are working hard to create molecular versions of the familiar electronic devices. Ways to make assemblies of just a few atoms behave like diodes, transistors, switches, wires and even crocodile clips are now under intense investigation. The potential for reduction is enormous. The powerful microprocessor found in the average desktop computer

carries several million microelectronic devices on each silicon chip. These devices are the logic gates and transistors that carry out unimaginable numbers of calculations every second. Molecules are at least a thousand times shorter than the smallest components etched in silicon, so several million million could fit on a molecular-based chip. That's the potential equivalent of a million desktop computers in a single machine.

EVER-DECREASING DESIGNERS

There are hundreds of research groups laying the foundations for molecular devices that might, one day, form the basis of this notional megaprocessor. One such group is that led by Stoddart. He and his colleagues have a deserved reputation as experts in designing molecules that undergo recognition and self-assembly, and they have made startling progress towards molecular devices that could act as switches in an entirely molecular computer.

Stoddart's team have come up with several variants on a similar theme: the molecular interlocking that, they hope, will allow them to understand the mechanics of recognition and self-assembly, and allow them to build molecular switches. They have perfected the art of threading ring molecules onto either a stopped molecular

K. ERIC DREXLER

Up until now attempts to design a molecular computer have simply been based on mimicking conventional computers on a minute scale. Drexler suggests that a completely different approach might be more successful.

The search is on for molecular 'crocodile clips' to link the molecular switches together

In molecular computers, the switches that will do the computing are individual molecules

LEFT **The problem with designing a molecular computer is wiring all the switching devices together. Current research is focusing on 'crocodile clips' to make the connections.**

string or another ring. Both the components of these supermolecules, known respectively as rotaxanes and catenanes, have inbuilt recognition capabilities. These play a crucial role in the self-assembly of the system, and provide the structure and information storage capacity that could allow it to behave as a binary switch.

The prototypical device has a positively charged macrocyclic bead with equal affinity for two recognition sites under normal conditions. The bead shuttles rapidly and precisely between the two sites several hundred times a second at room temperature. This kind of device will be of use only if the shuttling process can be controlled using an external signal. A laser light source could provide such a signal; output from one part of a molecular chip, it could be used to switch a molecular shuttling bead from one recognition site to the other in a shuttle. Such a flip of state could then represent a binary unit.

OPENING CHEMICAL GATES

A team led by Prasanna de Silva of the Queen's University, Belfast is working more directly on developing molecules that can perform the logical operations that are the basis of computations – logic gates. In a groundbreaking research programme, de Silva and his team have found that certain receptor molecules can actually perform simple computation-like reactions by coupling together two chemical processes – bonding between charged metals, ions, and molecular recognition processes.

In one example of such a device, the researchers have used a crown ether attached to an organic molecule known as anthracene. The molecule has two input channels depending on whether it binds sodium or hydrogen ions, or both, in solution. If both hydrogen and sodium ions are present above a threshold concentration, the molecule glows – a signal that can easily be detected.

Once the various research groups have devised ways to make the molecular logic gates, the next step is to piece them all together in a coordinated way. This would create arrays of devices on molecular chips, which could then be wired to an input device such as a mouse, keyboard or pen pad, and an output device such as an ordinary computer display screen.

Jean-Marie Lehn's team in Strasbourg is just one group working on the molecular wires that might interconnect switches like Stoddart's shuttles. They have prepared a molecular wire using a long-chain caroviologen molecule, and then studied the ease with which electrons flow between its terminals. According to their measurements electron conduction is possible in this supramolecular-scale system. Joining the wires to molecular devices remains a challenge, however, and research to devise molecular 'crocodile clips' is underway.

FLEXIBLE VISION

Scientists at the **University of Cambridge** and at the **Uniax Corporation** in California, as well as numerous research groups in Japan, Sweden and elsewhere, are working on materials that will form the next generation of computer displays. They are using a combination of innovative polymer chemistry and ingenious physics to build novel light-emitting diodes, which could eventually form the basis of large, flat, and even flexible displays. They believe these devices will consume far less power than the cathode ray tubes used in screens today, but be just as bright. On a smaller scale, such devices could become the optical signal sources that read and write to arrays of memory molecules. One such research group is led by physicist **Richard Friend** at the **Cavendish Laboratory** in Cambridge, England. Working closely with University chemists, they have made derivatives of electroluminescent polymers which can be sandwiched between a transparent layer and electrical contacts to make light-emitting diodes.

NANOCIRCUITS

In conventional electronics, it seems that silicon technology can only go so far. Microprocessor circuit printing techniques depend on the wavelength of the light source used, so X-rays represent the size limit for individual features. Attempts are being made to overcome this barrier, using self-assembling strands of DNA on silicon as a mask for chemical etching instead of photolithography.

Syracuse University researcher Robert Birge has come up with a way of using tiny molecular cubes of the microbial protein bacteriorhodopsin as a memory device. Information can be written to the protein and read from it by laser light. Theoretically such devices could hold 300 times as much data as modern memory chips. Groups of 'quantum dots', each containing a small adjustable number of electrons, might also find a place in this quest.

The research into such systems is being done by some brilliant scientists, so anything is possible, but it will take time. Molecular electronics cannot even be described as being in its infancy, for it is still in the embryonic stage, and molecular computation is just a twinkle in the eye of blue-sky researchers. But the next century could see a computer revolution that will turn Silicon Valley on its head.

MILLENNIAL CHEMISTRY

A chemist is an architect on the smallest possible scale: the molecular scale. Such molecular architecture is an enthralling process. A chemist may discover strange medicinally active compounds in rare South American orchids, extract them, isolate them and, by analytical cross-examination, identify them and their properties. He can then use all this information to make an exact copy of the molecule in question in the laboratory, not only for the sake of conservation but so that new, more effective and safer versions can be made and tested too.

Chemists are creating molecular machines, mimicking nature's nanometre-sized enzymes and receptors. A chemist can tailor such a device to precise requirements, enabling it to be used as a sensor to help diabetics measure their blood sugar, or a catalyst to speed up an industrial reaction. The sheer elegance of such solutions for medical, industrial and environmental problems is breathtaking when they are compared with the sledgehammer technologies we have relied upon in the past.

Chemistry may still have some way to go before it is allowed to shed its traditional image of the white coat and the test tube, but ultimately it may become the most exciting branch of science in the new millennium.

BELOW **The power of the computer has enabled chemists to work on a molecular level, building new substances and structures from the bottom, making chemistry the science most likely to revolutionize our lives in the next milennium. Even the computers themselves may be built by chemists.**

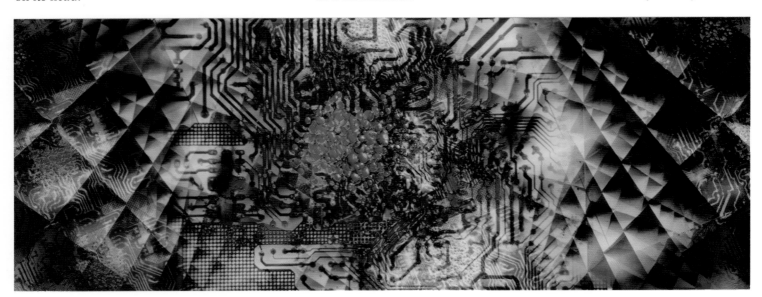

AT A GLANCE

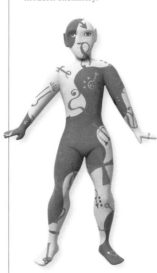

TRADITIONAL CHEMISTRY

1661

Robert Boyle writes *The Skeptical Chymist*, in which he introduces the notion of elements and compounds and lays the foundations for modern chemistry.

THE FOUR ELEMENTS

1756

Joseph Black recognizes the importance of recording weight changes and the role of gases in chemical reactions. He also deduces the presence of carbon dioxide in the atmosphere.

1774

Joseph Priestley discovers a wide range of new gases, including nitrogen dioxide, ammonia, nitrogen, carbon monoxide, sulphur dioxide and oxygen.

1800

Humphry Davy discovers nitrous oxide – laughing gas – and suggests its use as an anaesthetic.

1808

John Dalton proposes his atomic theory, suggesting that every element is made of minute particles called atoms which can be neither divided nor destroyed. According to his theory, every atom of each element is identical.

JOHN DALTON

1810

Discovery of supercritical fluid – a state between liquid and gas.

1811

Amadeo Avogadro formulates Avogadro's Law, which provides a basis for working out the chemical formulae of gases.

CHEMICAL REACTIONS

1828

Friedrich Wöhler shows that organic chemicals are not unique to living things by synthesizing urea, pioneering the growth of organic chemistry.

1850

Louis Pasteur discovers chirality – mirror-image molecules.

1868

Dmitri Mendeléev creates the Periodic Table of chemical elements, in which the elements are placed (initially) in order of increasing relative atomic mass, and arranged into vertical groups possessing similar chemical characteristics.

1874

Jacobus Van't Hoff pioneers the idea of stereo chemistry from Pasteur's discovery of chirality.

EMIL FISCHER

1899

Emil Fischer works out the structure of sugar, and synthesizes glucose and fructose – a fundamental breakthrough in organic chemistry.

THE ATOM

1905

Paul Ehrlich develops the idea of the magic bullet – a chemical which can find and destroy a pathogen.

MOLECULAR COMPUTERS

1962

Elias Corey develops the idea of retrosynthetic analysis for building large, complex organic molecules from smaller ones.

1966

Charles Pedersen designs crown ethers – the first chemical compounds able to recognize others.

1980s

Jean-Marie Lehn, Donald Cram and Charles Pedersen pioneer designed molecular complexes.

1985

Robert Curl, Sir Harold Kroto and Richard Smalley discover buckyballs.

BUCKYBALL

1994

Groups led by Kyriacos Nicolau and Robert Holton succeed in synthesizing the natural cancer treatment compound Taxol.

1995

John Heckels and Myron Christodoulides develop experimental peptides that provide protection against Hepatitis-B.

VIRTUAL CHEMISTRY

The Dynamic Earth

EARTH SCIENCES

The Earth sciences, more than any other field of scientific enquiry, are inextricably linked to the passage of time. The study of geology – the foundation of all Earth sciences – has been described as the science of telling time, and establishing the dates and sequences of events on Earth. Sooner or later any geological investigation moves from process to history, and to events that occurred during a particular period in the past. So measuring time – to a geologist – is almost a religion.

ABOVE **If it is found preserved as a fossil on a mountain, far from any ocean, a simple seashell can have astonishing implications for the history of the Earth.**

No other field of science has found it necessary to construct its own special scale of time. There is no formalized biological timescale, or chemical timescale. All other fields of study simply use the intervals of time known to us all: seconds, minutes, hours, days, years. Geologists, on the other hand, talk about periods and epochs, eras and zones, stages and series: the subdivisions of what is known as the geological timescale. The measurement of time is clearly indispensable to the study of geology, and it is also a source of great error, confusion and even heartbreak for all geologists and other earth scientists trying to wrest the truth from our planet's rocky cover.

BELOW **Titanic forces, operating over immense periods of time far beyond our own limited human experience, have created landscapes of sublime beauty and, to an earth scientist, endless fascination.**

INDUSTRY AND SCIENCE

Geology, one of the key Earth sciences, has ranked as a modern scholarly discipline for only two centuries. It was created as much for economic reasons as for scientific ones, and this uneasy alliance between economic and scientific motives has had a marked effect on its history. It was born out of necessity, and is really a scientific child of the European Industrial Revolution. There would be no coal, gas, iron, or a thousand other fuels and minerals without geological investigation, and just as nineteenth-century Europe needed these materials to run its emerging industries, our society needs them still.

Yet geology is not simply a research arm of industry. It has become a vibrant field of intellectual enquiry, and is now merging with other sciences to examine scientific questions far beyond its original domain.

For example, geology and astronomy now jointly peer into the heavens at distant planets, and geology has joined forces with biology to help unlock the mysteries of life. Some of the most exciting scientific discoveries of any scientific discipline have come from geology. The discovery that continents can drift across the surface of the planet, our under-standing of the age of the Earth, and the unravelling of the history of life can all be attributed to geologists.

MYTH AND TRADITION

Geologists became concerned with time for two quite different reasons. The first was economic. Early geologists found that they could find economically valuable minerals and fuels more easily if they could understand the structure of the Earth's surface. They quickly realized that, to do this, they needed some way to date rocks. The second reason, however, was more academic: many scientists were mighty curious about how many thousands or millions of years old the Earth is.

Finding the age of the Earth was a quest that had long been entrusted to theologians, who searched not in the structures of the Earth itself, but among the sacred writings of human prophets. Their answers varied between a few thousand years and infinity. The Hindu tradition weighed in at slightly less than two billion years, while some Hebrew and Christian calculations resulted in values of less than 10,000 years. Initially the discrepancies didn't really matter, for the question was of no practical importance. But with the onset of the Industrial Revolution in the eighteenth century, and the resulting thirst for metals and fuels, all that changed.

ABOVE Early estimates of the age of the Earth were derived from creation myths. In the seventeenth century a careful study of Genesis led James Usher, archbishop of Armagh, to conclude that the world was created on 23 October 4004 BC, at nine o'clock in the morning.

LEFT Fuelled by coal and wrought from iron, the Industrial Revolution heralded a new era of hunger for natural resources. Geology suddenly acquired vital economic importance, and started developing into a real science.

THE APPLICATION OF SCIENCE

The first people to grapple with scientific approaches to understanding the age of any rock – and therefore the age of the Earth – were engineers working in the mining regions, who needed a better system of understanding the apparently chaotic piles of rock forming the surface of the Earth. They gradually realized that if the various rock units could be dated by their relative ages, a correlation could be established between rocks that might lie some distance apart. In this way, some order could be recognized among the geological chaos. But how could rocks be accurately dated?

Eighteenth-century European geologists believed that a rock's type was the best clue to its age. They assumed that all rocks of a specific type, such as all granites, had been formed at the same time. By the early 1800s, however, they had come to believe, correctly, that the composition or mineral content of a rock is virtually independent of its age.

With their main dating tool discredited, geologists were in despair. But then a new way of finding the age of at least some types of rock was found: the use of the strange, petrified remains of animals and plants that were often to be found in certain rocks: fossils.

BELOW **The spectacular Grand Canyon of Arizona has been created by the Colorado River cutting down through horizontal layers of rock, laid one upon the other. The deeper the layers, the older they are, and the rocks at the very bottom of the Canyon are two billion years old.**

FOSSIL EVIDENCE

When animals and plants die their remains are usually destroyed, but sometimes they are buried in the mud or other sediments while they are still fairly complete. The softest parts usually rot away quite soon, but the harder parts such as shells or bones survive for much longer. If the sediments surrounding them harden into solid rock these preserved remains may survive for millions of years as fossils.

Many fossils are of immense interest because they are the remains of organisms that have become extinct. When the rocks that yield them can be dated, either absolutely or relatively, the fossils show how life has changed over time, or evolved. But fossils can also return the compliment.

Since most species of animals and plants exist on Earth for only a relatively short time – geologically speaking – they can define the ages of the rocks that contain them. Their actual nature is irrelevant for this purpose, so long as they can be identified. So although William Smith – the man who first used fossils in this way – probably knew very little biology and nothing about evolution, he could classify rock strata by the identifiable fossils they contained. He could also use this classification to devise the first geological map.

FOSSIL CLUES

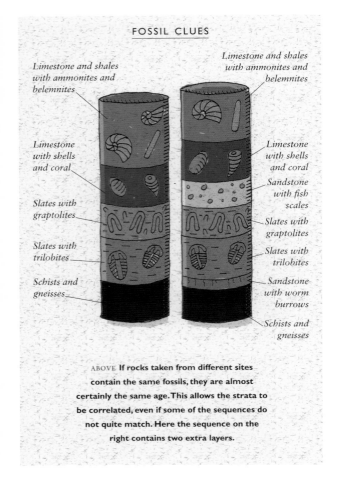

Limestone and shales with ammonites and belemnites

Limestone and shales with ammonites and belemnites

Limestone with shells and coral

Limestone with shells and coral

Sandstone with fish scales

Slates with graptolites

Slates with graptolites

Slates with trilobites

Slates with trilobites

Sandstone with worm burrows

Schists and gneisses

Schists and gneisses

ABOVE **If rocks taken from different sites contain the same fossils, they are almost certainly the same age. This allows the strata to be correlated, even if some of the sequences do not quite match. Here the sequence on the right contains two extra layers.**

A POWERFUL NEW TOOL

Fossils had long been known to science, but until the close of the eighteenth century they had never been viewed as anything other than curiosities, with no scientific value. But a very practical Englishman, armed only with hammer and chisel, changed that view forever.

William Smith was a surveyor employed by the engineers building Britain's expanding system of canals in the last years of the eighteenth century, and he became intrigued by the fossils he routinely encountered in the course of his work. As he examined the cuttings excavated along the canal routes he realized that he was seeing the same succession of fossil types in the rocks.

Many naturalists before Smith had recognized that fossils from a lower succession of rock layers, or strata, were often different from the fossils found in younger, overlying strata. But prior to Smith no-one had recognized that the succession of fossils in strata was often the same from region to region. Here was the stroke of genius: it was not any individual fossil which determined age, but the characteristic succession of many fossils, and the fact that the same groups of fossils occur in the same sequence wherever they are found in a given region. This allowed Smith to correlate strata in rocks that were some distance apart, and not obviously related.

At first, Smith's great discovery was known only to his own circle of acquaintances. Indeed, he first announced it in a pub. But gradually word of this new system spread, and by the second decade of the nineteenth century geologists began to realize that specific types of rocks could be formed at any time, but specific fossils were formed only at particular times in the history of the Earth.

Here, then, was a powerful new tool. It could not determine the actual age of a rock in years. But it could be used to make a very accurate determination of which rock was younger and which was older – in other words, their relative age. For most practical purposes, this was enough. It enabled the structure of the Earth's surface to be understood and mapped, and this led in turn to the discovery of its underground secrets and mineral treasures.

BELOW **William Smith created his geological maps by painting the information onto existing maps published by John Cary of London, using watercolours to indicate different rock strata. This colour-coding convention is still used on modern geological maps.**

BELOW **Having created his maps of rock outcrops, Smith was able to deduce the arrangement of strata beneath the ground, along a line extending from London to the mountains of North Wales.**

HIDDEN LAYERS

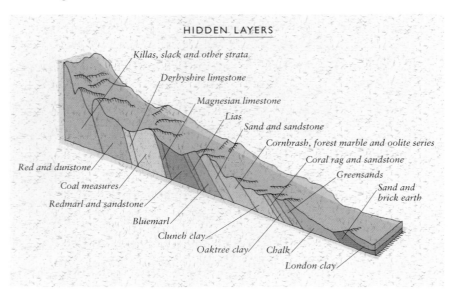

Killas, slack and other strata

Derbyshire limestone

Magnesian limestone

Lias

Sand and sandstone

Cornbrash, forest marble and oolite series

Coral rag and sandstone

Greensands

Sand and brick earth

Red and dunstone

Coal measures

Redmarl and sandstone

Bluemarl

Clunch clay

Oaktree clay

Chalk

London clay

STRATA AND THE PRINCIPLE OF SUPERPOSITION

One of the basic principles of geology is the principle of superposition, which was first recognized in the seventeenth century by a Dane called Nicolaus Steno.

In any area where rocks are laid down in layers, known as strata, then common sense decrees that the lowest strata will have been laid down first, and the others laid on top. This basic principle holds good regardless of whether the rock strata are formed from solidified mud or other sediments, or represent successive flows of lava, ash or other material that has been ejected from volcanoes and spread over the surrounding landscape.

Where rock strata form neat horizontal layers the principle can be applied with some confidence. But sometimes the layers have been buckled, folded and faulted by the forces that cause earthquakes, and they may even be overturned. Molten rock can also be intruded between layers of older rock to disrupt the sequence. So judging the relative age of rocks using the principle of superposition has its pitfalls. Luckily there are ways of avoiding them, and the principle does have one great advantage: it can be applied in the field without any special equipment other than a well-trained eye.

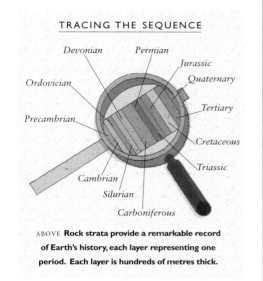

TRACING THE SEQUENCE

Devonian · Permian · Jurassic · Quaternary · Tertiary · Cretaceous · Triassic · Carboniferous · Silurian · Cambrian · Precambrian · Ordovician

ABOVE **Rock strata provide a remarkable record of Earth's history, each layer representing one period. Each layer is hundreds of metres thick.**

THE GEOLOGICAL TIMESCALE

The technique of using fossils as relative indicators of time was soon taken up by geologists all over Europe. Major divisions of time were defined on the basis of their unique fossil assemblages. These divisions became the units of geological time.

As the stratigraphic record – the information preserved in rock strata – was examined more closely, geologists became aware of two points where the changes in fossil assemblages were especially dramatic. At the top of the sequence of strata named the Permian system, and at the top of a much younger sequence of strata known as the Cretaceous system, the vast majority of animal and plant fossils were replaced by radically different types of fossils. Nowhere else in the stratigraphic record were such abrupt and all-encompassing changes to be found – changes that pointed to wholesale extinctions of the plants and animals represented by the fossils found in the rocks.

These two radical turnovers in the fossil record were so dramatic that they were used by an Englishman named John Phillips to subdivide the stratigraphic record – and the history of life it contains – into three large-scale blocks of time: the Paleozoic, Mesozoic and Cenozoic Eras.

The Paleozoic Era, or time of 'old life', extended from the first appearance of creatures with skeletons 530 million years ago to a major extinction 250 million years ago. The Mesozoic Era, or time of 'middle life', began immediately after the great Paleozoic extinction and ended 65 million years ago with the extinction of the dinosaurs. The Cenozoic Era, or time of 'new life', has extended from the last great mass extinction to the present day.

LEFT **For much of the Mesozoic era, which began 250 million years ago, the Earth was dominated by the dinosaurs. The Mesozoic ended with their extinction 65 million years ago.**

WILLIAM SMITH
1769–1839
An engineer and surveyor, Smith produced the first proper geological map by correlating fossil assemblages in widely separated outcrops of rock strata.

SEDIMENTARY, METAMORPHIC AND IGNEOUS ROCKS

Fossils are found in sedimentary rocks, which are hardened and cemented layers of mud, sand, lime particles or other granular materials. Most sedimentary rocks such as limestones and shales were formed from beds laid down underwater, although many sandstones started life as sand dunes and other rocks originated as piles of debris dumped by glaciers.

If these rocks are subjected to extreme heat or pressure by the titanic forces that build mountains, they can be turned into different types of rock in a process called metamorphism. So if limestone, for example, is heated enough it can turn into marble, and if shale comes under enough pressure it can turn into slate.

If one of these metamorphic rocks is heated or compressed even further, it may change into yet another type of metamorphic rock. Slate can become phyllite, then schist, then ultimately gneiss as the forces that cause metamorphism become increasingly intense. Not surprisingly, these processes normally destroy any fossils that may have been preserved in the original sediments.

The heat that often causes metamorphism is usually generated by molten rock welling up from below. When this molten rock solidifies it forms igneous rocks such as granite and basalt. These rocks are generally very hard, because they are composed of interlocking crystals of minerals such as quartz and feldspar, olivine and pyroxene. When they are exposed to the weather, though, they gradually break down into fragments which are carried away by rivers and washed into the the sea – where they are laid down in the beds that form sedimentary rocks.

THE ROCK HIERARCHY

Phillips suffered no misapprehensions about the cause of these three divisions: he realized that twice in the Earth's past, life had been virtually extinguished, and that only a tiny fraction of species had survived the catastrophes, whatever their cause. Phillips concluded

that the history of life on Earth had, in the past, been interrupted by periods of wholesale destruction that we now call mass extinctions.

In this way a hierarchy of rock units was established, each based on actual rocks and the fossils they contained, but with no dates assigned to them. The Jurassic system, for example, was defined as all the sedimentary rocks deposited between the first occurrence of one species of ammonite fossil, and the last occurrence of another species of ammonite fossil which clearly lived much later. The period

during which the Jurassic system was being deposited was defined as the Jurassic Period, but since no-one knew when the Period started or when it finished, no-one knew how long it lasted.

So telling time using fossils has severe limitations. What's more, assigning even a relative age was impossible for rocks that did not contain fossils, such as the igneous rocks formed from solidified lava, or metamorphic rocks subjected to extremes of heat or pressure. Even in sedimentary rocks fossils are only occasionally abundant, and many sedimentary rocks, such as those deposited

by glaciers or rivers, never have fossils because the remains are destroyed before they can be preserved. As a result, some other method had to be found.

ABOVE AND LEFT **Fossils, the petrified remnants of long dead organisms, provided the first reliable way of assessing the relative ages of rock strata. Identifying key fossils which only survive at a particular time in particular environments gives a fairly clear indication of how the rock fits into the time sequence. But fossils are only found in sedimentary rocks – rocks that form from sediments on the sea bed – and only provide a relative indication of the rock's age, not its absolute age.**

KELVIN'S COOLING EARTH

By the late 1800s, the search for a more reliable means of dating non-sedimentary rocks was a major goal of Earth science. Economic motives were still important, just as they had been at the start of the nineteenth century. But as the century came to a close, scientists became more interested in determining the age of the Earth itself.

Two schools of thought emerged. Most geologists, recognizing the immense thickness of sedimentary rocks found over the globe, believed that billions of years had passed since the Earth was formed. The great diversity of fossils encased in all this strata seemed to support this view, since according to Darwin's theory of plant and animal evolution by natural selection, immense periods of time were necessary to permit the evolutionary changes needed to create the great diversity of living (and extinct) organisms. On the other hand, physicists thought the Earth was, at most, a few hundred million years old, and the arguments between the physicists and the 'naturalists' were hard-fought and often extremely bitter.

The man who most famously studied the question in the nineteenth century was William Thomson – better known by his title Baron Kelvin of Largs, or Lord Kelvin for short. Through brilliant insight and mathematical calculation, Kelvin made the first scientific computation of the age of the Earth: less than 100 million years.

EVOLUTION AND TIME

Darwin's theory of evolution by natural selection provides the scientific foundation for the most common method of determining geological time: the practice called biostratigraphy. Essentially, different forms of animals and plants evolve through time in a recognizable sequence, and by identifying the fossil remains of these biological forms preserved in the rock strata, it is possible for the rocks to be dated.

When biostratigraphy was first applied, at the end of the eighteenth century, it was used to assign relative dates to the rocks. No-one knew how old the fossils were in absolute terms, or how many years had elapsed between the deposition of the different fossil forms. But when Darwin's theory of natural selection – first published in 1858 – became generally accepted, it became apparent that evolution is an extremely slow, largely random process, in which new forms emerge by accident, and usually disappear before they can reproduce themselves. The time taken for a new type of organism to establish itself and appear in the fossil record had to be reckoned in millions, rather than thousands of years.

So by simple arithmetic it was clear – to the evolutionists – that the huge variety of organisms on Earth must have evolved over a vast span of time. This meant that the planet had to be far older than the 100 million years calculated by Kelvin and the other physicists.

CHARLES DARWIN
1809–82
Through a combination of inspiration and painstaking research, Darwin came up with a mechanism for evolution that could only operate over immensely long spans of time, forcing geologists to revise their estimates of the Earth's age.

Fossil dating
Each period of geological activity has left its own particular range of fossil species – so we can date a rock from the kind of fossils it contains

Since fossils occur only in sedimentary rocks, and only in rocks laid down since the evolution of recognizable life forms, they cannot be used to date igneous or ancient metamorphic rocks

Radioactive dating
All rocks are built up from minerals, which are compounds of elements. Each element is composed of atoms with nuclei built up from many nucleons

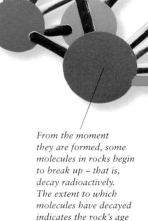

From the moment they are formed, some molecules in rocks begin to break up – that is, decay radioactively. The extent to which molecules have decayed indicates the rock's age

Kelvin's estimate was based on the study of heat and heat flow. By Kelvin's time it was known that each translation of energy from one form to another resulted in the formation of heat, which is then dissipated. This fact, which eventually become known as the second law of thermo-dynamics, is the reason why perpetual-motion machines are impossible. Kelvin reasoned that the Earth and Sun are both cooling, and that by establishing the initial temperatures of each of these bodies as well as their rate of heat loss, he could arrive at an estimate of their age.

Kelvin's method for working out the age of the Earth involved finding the rate at which heat was lost by both the Sun and the Earth. While measuring the rate of heat loss from the Sun was difficult, measuring the heat loss from the Earth was relatively easy. He needed three values: the initial temperature of the Earth, the thermal conductivity of rocks, and the actual heat flow. He used a value of 3,870°C as the initial temperature of the Earth, since this was thought to be a reasonable estimate for the temperature at which rocks melt to a molten state. In doing so, he assumed that the Earth had formed from a hot, molten mass rather than coalescing from many cold fragments. He obtained heat flow measurements for various rock types in the laboratory. All that was wanting was a measure of the actual rate at which heat was escaping from the Earth's interior, and he found this by measuring the Earth's temperature at various depths in underground mines.

These experiments showed that heat is indeed flowing outward from the deep interior of the Earth – so much heat, indeed, that the Earth must have formed less than 100 million years ago. What Kelvin failed to appreciate – quite reasonably – was the fact that heat is still being generated inside the Earth, so it is cooling down far more slowly than his heat-flow data suggested.

Paleomagnetic dating
Taking a core sample of a rock and recording its precise orientation enables the rock to be dated by paleomagnetism

Tiny magnetic crystals within many rocks are aligned with the Earth's magnetic field, as it was when the rock solidified

Each radioactive element decays at a constant rate to form a daughter element, so counting the proportions of parent and daughter atoms in a rock gives a measure of its age

Certain minerals within rocks turn like a compass to align with Earth's magnetic field at the time they were formed. The direction can be compared with known polarity changes in Earth's magnetic history to establish the age of the rock

ACCUMULATING ROCKS

One of the main objections to the 'young Earth' hypothesis was the sheer thickness of sedimentary rocks to be found on the Earth's surface. In 1895 geologist William Sollas estimated that if the sedimentary rocks of different ages were to be piled together in one continuous column, the column would be 100km tall – which is more than 11 times the height of Mount Everest.

Such a column would certainly take some time to form, assuming that the processes (and rates) of sedimentary rock accumulation have been uniform throughout time. This assumption was justified by the principle of uniformitarianism established by James Hutton (1726–97). Often considered the 'father of geology', Hutton realized that the rocks of his native Scotland were the results of processes that could be observed in action all around us – processes such as erosion, deposition and volcanic eruptions – and that these processes operated at the same rates regardless of when they occurred.

Sollas asked the following question: what is the rate at which sedimentary rocks, past and present, can be expected to accumulate? Assuming an accumulation rate of about one metre every 300 years, he arrived at an estimate of about 34 million years for the age of the Earth since sedimentation began. However, this was very much a minimum figure, for it assumed that the layers of sedimentary rocks had accumulated without pause or break. This is not the case in nature, because sedimentation is interrupted by periods of erosion. Furthermore, Sollas' estimate did not account for the long period of time before sedimentation began, when the Earth's surface was still in a molten state. In fact the oldest sedimentary rocks so far discovered were formed over 3,400 million years ago – 100 times Sollas' estimate.

ABOVE **Kelvin assumed that the lava erupting from volcanoes was molten rock left over from the formation of the Earth. In fact it is created by the heat of radioactivity within the Earth – a source of energy that was still unknown in Kelvin's day.**

RIGHT **As atoms of radioactive uranium-235 lose nucleons they are transformed into lead atoms. Half the remaining uranium turns to lead every 704 million years, so if a rock sample contains uranium-235, measuring the proportions of uranium and lead shows how many 'half-lives' have elapsed since the rock was formed.**

RADIOACTIVE HALF-LIFE

Some of the elements forming the rocks of the Earth are unstable, because their atomic nucleii spontaneously fragment into smaller nucleii, releasing energy in the process. This phenomenon is called radio-active decay. When an atom loses some of its nuclear mass in this way it becomes another element, so when an atom of uranium-235, which has 235 'nucleons' in its nucleus, loses 28 of them through radioactive decay, it become lead-207. The decay process occurs at a steady rate, and it has been found that half the atoms of any sample of uranium-235 will decay to lead-207 in 704 million years. This is called its radioactive half-life.

This means that a rock containing uranium 235 can be dated by measuring the number of uranium 'parent atoms' and lead 'daughter atoms'. Comparing the two gives an estimate of the rock's age, because if, say, there are three times as many lead atoms as uranium atoms, then the rock is two half-lives old, or 1408 million years. Other radioactive elements decay at different rates, but they can be dated using the same principle.

RADIOACTIVE DECAY

As uranium loses atoms it turns into lead

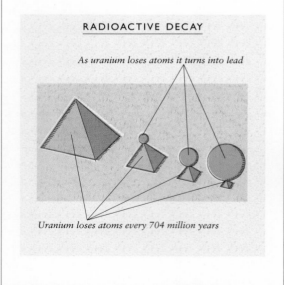

Uranium loses atoms every 704 million years

Kelvin's calculations were considered highly authoritative, not so much for how they were done, but because of who Lord Kelvin was. His conclusion about a relatively young Earth was soon supported by further work dealing with heat flow, this time by Clarence King, Director of the Geological Survey of Canada. King, using Kelvin's methodology, made new estimates of heat flow and rock conductivity. King's new estimate was even shorter than Kelvin's: he calculated that the Earth was only 24 million years old.

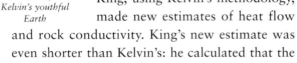

24,000,000 years
Kelvin's youthful Earth

The Kelvin-King conclusions on the age of the Earth were widely, but not universally accepted. The chief dissenters were those familiar with evolution and the fossil record – such as Darwin's chief ally Thomas Huxley – and those who studied the rate at which sedimentary rocks accumulate. Twenty-four million years was far too short a time to accommodate all the evolutionary changes visible in the fossil record, and also far too short to account for the amount of sedimentary rock present on the Earth's surface.

100,000,000 years
Huxley's Earth was a little older

Another ingenious attempt at making a reliable estimate of the Earth's age came to be known as the 'salt clock'. The amount of salt in the ocean, and the rate at which it arrived there, was thought to be a way of calculating the age of the Earth. Several chemists arrived at an estimate of about 100 million years using this method, and by the twentieth century most scientists had come up with a figure of between 25 and 100 million years for the age of the Earth. Only the evolutionists found this implausible and, as we now know, their misgivings were well-founded. The true age of the Earth is measured in billions, not millions of years.

1,000,000,000 plus years
The true age of the Earth

RADIOMETRIC DATING

Towards the close of the nineteenth century, at the peak of Kelvin's eminence, the seeds of his downfall – at least as far as an estimate of the Earth's age was concerned – had just been discovered. In 1895 Wilhelm Roentgen of Germany discovered X-rays, while a year later Antoine Henri Becquerel of France discovered the strange properties of uranium. Two years after that, Marie Curie discovered similar properties for the element thorium, and coined the word 'radioactivity' to describe them.

Radioactivity was to become the key to accurate determination of rock age. The breakthrough was made by the physicists Ernest Rutherford and Frederick Soddy who, working at McGill University in Montreal, discovered the principles of radioactive decay and 'half-life'. In 1907 Rutherford made the first suggestion that radioactive decay processes could be used as a geological timekeeper, stating,

If the rate of production of helium from known weights of the different radio-elements were experimentally known, it would thus be possible to determine the interval required for the production of the amount of helium observed in radioactive minerals, or, in other words, to determine the age of the mineral.

With these prophetic words, a new era of age determination had opened. Rutherford calculated the ages of two minerals, finding minimum ages of 500 million years for each. Soon other workers began analyzing rocks containing radioactive elements, using slightly improved methods that yielded a series of ages all well over the 100 million years calculated by Kelvin as the maximum age of the Earth. The oldest of these was 2.2 billion years. The discovery of radioactivity – and the natural clock which it keeps – made Kelvin's simple conductive calculation obsolete.

But why was Kelvin's result so different from the ages found by the pioneering geochronologists? Kelvin's estimate was based on his assumption that the Earth was simply cooling down from a much hotter earlier stage, and the only other source of heat was the energy radiated by the Sun. We now know that the Earth is actually producing heat by the process of radioactive decay within its rocks – the very process that enables rocks to be dated.

The heat from this vast natural nuclear reactor prevents the Earth cooling down at the rate predicted by Kelvin, and gave him the impression that the whole process had started relatively recently. His calculations were correct, but they failed to take account of a phenomenon that, at the time, was unknown to science.

LEFT **The discovery of X-rays and radioactivity revolutionized our understanding of the Earth, providing a tool for finding the ages of rocks, and explaining the source of its internal energy.**

LORD KELVIN
1824–1907
As virtual founder of the science of thermodynamics, Kelvin used his understanding of heat flow – the dissipation of heat from the cooling planet – to calculate the age of the Earth.

BELOW **Kelvin's estimate of the Earth's age depended on working out the rate heat was lost by both the Sun and Earth. But the Earth's heat comes not only from the Sun's core but also its own interior radio-activity, so he underestimated just how old the Earth is.**

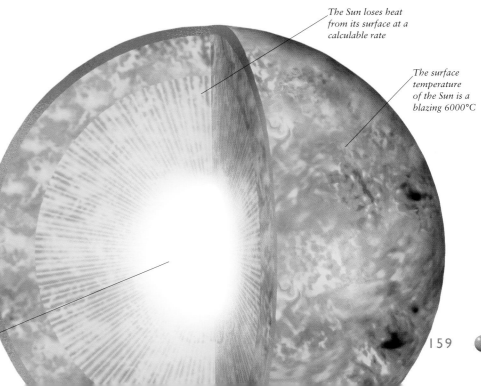

The Sun loses heat from its surface at a calculable rate

The surface temperature of the Sun is a blazing 6000°C

Heat is continually generated by nuclear reactions in the Sun's core

VOLCANIC CLOCKS

The early geochronologists examined the breakdown of uranium to lead, because uranium was the most strongly radioactive element. Today, other radioactive elements and their breakdown products are often used, allowing a wider variety of rocks to be dated. One of the most useful is a form of potassium, which decays to the rare gas argon. The quantities of both 'parent' and 'daughter' atoms in a rock sample are measured using a machine called a mass spectrometer. Today these are among the most common major laboratory instruments, with thousands operating daily throughout the world. Measurements made using such equipment show that the Earth is at least 4.5 billion years old.

Mass spectrometers have allowed us to date many types of rocks with high precision. Ironically, the technique works best on those rocks which most stubbornly resisted nineteenth-century efforts to determine ages – the igneous rocks that make up the cores of mountain ranges and the lava flows of the world's volcanoes. Radiometric dating works least well on those rocks containing fossils, and in fact the most sophisticated mass spectrometer in the world is useless on an ordinary fossil-bearing sedimentary rock, because the grains from which the rock is made are usually derived from much older igneous rocks.

Initially this frustrated all attempts to put absolute dates on the fossil-defined units of the geological timescale. The only solution was to find places where volcanic ashes or lava flows were interbedded with ancient sedimentary rocks. These became the new holy grail for Earth scientists. They learned to recognize a rock type called bentonite: a thin layer of orange volcanic ash which can be radiometrically dated, yet which commonly occurs in fossil-bearing beds. By dating bentonite bedded within a whole range of sedimentary rocks, geologists were finally able to integrate fossil and radiometric dating techniques.

BELOW Much of our knowledge of the inside of the Earth comes from studying the echoes of earthquake waves as they reverberate through the Earth. It is from these that we know that the Earth's core is likely to be made of the dense metal iron. It is the swirling of this hot liquid core which gives Earth its magnetic field.

POLARITY REVERSALS

Why polarity reversals take place is a mystery. There are many theories, and they all deal with the complex interactions taking place deep within the Earth. The Earth's core has never been observed and never can be, yet we know a great deal about it from the way earthquake waves move through the Earth. Unlike the overlying mantle the core – or its outer part at any rate – is liquid, and it is mainly made up of metallic iron and nickel. The core is the originator of the magnetic field, and its perturbations, eddies, convection currents and other movements seem to trigger the magnetic reversals. Perhaps periodic irregularities or interactions between the liquid core and the surrounding, solid mantle region somehow trigger the phase changes. Given the current state of our knowledge, we can only speculate.

MAGNETIC CLOCKS

The two most commonly-used dating methods in geology are identifying reliably-dated fossils and measuring the radioactive decay of one element into another. Yet if you live in a land where the fossils are unique and new, or where none of the rocks contain the elusive minerals with radiometric clocks ticking away within them, you have a problem. Luckily, other methods have been developed. One of the most powerful natural timekeepers is the Earth's magnetic field.

All magnets show the curious property of being bipolar – they have positive and negative poles. If the positive pole of one magnet is placed near the negative pole of another, they attract each other. But if you turn one of the magnets around so the adjacent poles are of the same type, they repel each other.

Permanent magnets always have the same polarity: their positive and negative poles are always in the same place. But an electromagnet, which has a magnetic field set up by an electric current flowing through a coil, changes its polarity if the electrical connections are reversed.

The Earth acts as if it has a huge permanent bar magnet in its interior, aligned roughly north-south. On the face of it, the Earth would seem to be an unlikely magnet. Why does it have magnetic properties at all? There clearly isn't enough metal in the crust to produce the enormous magnetic field surrounding the Earth – a field so strong that it can deflect solar radiation. It turns out that the field emanates from deep within the Earth, from a liquid iron core which behaves like a dynamo, generating an electromagnetic field. And like all electromagnets, it can change its polarity.

BELOW **The iron core of the Earth is largely liquid, and as it spins the swirling, magnetized liquid iron seems to generate looping electric currents that set up an electromagnetic field around the planet. Occasionally disruptions in the system put the mechanism into reverse, so the field switches polarity.**

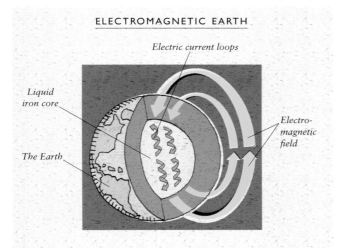

ELECTROMAGNETIC EARTH

Electric current loops

Liquid iron core

The Earth

Electro-magnetic field

RIGHT **It has been discovered that the Earth's magnetic field has completely reversed its polarity every few hundred thousand years, with the north pole switching to the south and vice versa. The record of these reversals provides a time scale by which rocks can be dated.**

FOSSIL COMPASSES

When molten rock solidifies, it preserves a record of the Earth's magnetic field at the time of solidification. In the early 1900s two French geologists, using the most primitive of equipment, discovered that the same lava outcrops in France preserved two diametrically opposed directions of polarity. Since the rocks in which the polarity was detected had long since solidified, they were acting rather like fossil compasses, indicating the directions of the Earth's magnetic poles at the time of their formation. The fact that two opposing directions were found in the same masses of lava created a debate which continued for decades. Finally, after repeated measurements of this 'paleomagnetism' with increasingly sophisticated equipment, there was but one inescapable conclusion: the Earth's magnetic field had somehow switched polarity. The rotation and the orientation of the Earth have not changed, but the direction of its magnetic field has.

Scientists had known about polarity reversal for 50 years before its implications for calibrating geological time became apparent. In the 1960s geologists began sampling thick lava flows on the edges of volcanoes. Because each individual flow could be dated using potassium-argon techniques, they were able to record a series of ages for the flows. Each dated flow was then sampled for its paleomagnetic direction. To the surprise of the investigators, they found not only both normal and reversed magnetism, but a whole series of reversals, and each could be radiometrically dated.

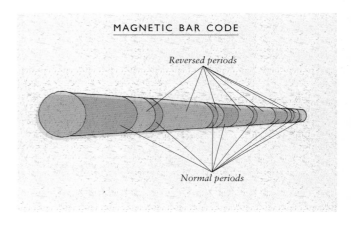

MAGNETIC BAR CODE

Reversed periods

Normal periods

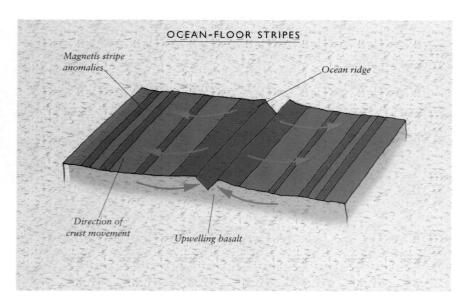

OCEAN-FLOOR STRIPES

Magnetis stripe anomalies

Ocean ridge

Direction of crust movement

Upwelling basalt

ABOVE **Molten basalt welling up at a mid-ocean ridge freezes solid when it contacts the cold water, and locks into its structure a record of the Earth's magnetic field at that time. Over the years, movements in the mantle beneath the crust drag the ocean floor away from the ridge on each side, preserving the magnetic data in a mirrored series of ocean-floor 'magnetic stripe anomalies'.**

OCEAN-FLOOR STRIPES

The geologists soon realized that the present, 'normal' magnetic field direction has existed for only about the last half million years. Prior to that, the field was 'reversed', and before that it was normal again. As they sampled ever-older piles of lava, they learned that the interval between reversal episodes ranged from hundreds of thousands to millions of years.

At about the same time, oceanographers made a similar discovery for underwater volcanoes. In the early 1960s it was discovered that new oceanic crust is created from basalt lava, extruded by elongated volcanoes arranged along submarine mountain chains called spreading centres. The new ocean crust is then carried away from the spreading centre on each side. When oceanographers towed magnetometers over the spreading centres, they found stripes of normal and reversed polarity symmetrically arrayed around the spreading centres.

Since they knew that the stripes of oceanic crust were formed sequentially, all they needed to obtain a chronology of the reversals were dates for each stripe. These were obtained by a ship designed to drill for core samples of rocks on the sea bed, and by the late 1960s these cores had provided enough information to enable a 'Geomagnetic Polarity Timescale' to be constructed.

By themselves, magnetic reversals are virtually useless – they are simply positive or negative binary data, like the data employed in a computer. But just as a simple plus and minus computer code can record or reveal great quantities of information when enough data is accumulated, so too can the binary code of reversal history. If it is combined with other techniques such as radiometric dating, the pattern becomes a very powerful tool.

At first the code could not be easily related to sedimentary rocks. What was needed was a readily accessible record of hard, magnetically measurable strata containing both fossils and a paleomagnetic signal. If a thick sequence of such strata could be found, the fossils could be used to assign dates to the pattern of reversals. Such a record was discovered in the early 1970s, in the Appennine Mountains of Italy.

A MAGNIFICENT RECORD

The discovery was made by geologist Walter Alvarez and his co-worker Bill Lowrie, more or less by accident. They were sampling magnetically-oriented cores from thick successions of white limestone, looking for evidence that the Italian peninsula had 'rotated' in the past owing to movement of the plates forming the Earth's crust. Laboratory analysis of these cores showed that there was so much movement of the various limestone layers that a detailed history of plate motion could not be obtained. What was available, however, was an unexpected and magnificent record of magnetic reversals.

Because these Italian limestones were rich in microfossils, (mainly the skeletons of single-celled planktonic organisms called foraminiferans), the individual magnetic reversals could be correlated to those of radiometrically dated deep-sea lavas. For the first time, European fossils could be linked to radiometric dates by using the magnetic reversal history as a go-between, and it was confirmed that the detailed, continuous deep-sea record of magnetic field polarity change could also be detected in sedimentary rocks found on land.

GEOMAGNETIC EVIDENCE

The evidence for geomagnetic reversals comes from the orientations of microscopic magnetized mineral particles locked within either sediments or lava. The most common of these magnetic minerals, named magnetite, forms rod-shaped crystals, each with a positive and negative pole like any other magnet. If these mineral grains exist in a medium where they can freely move, such as water or liquid lava, they act as miniature compass needles. The positive poles of these magnetized grains point toward the negative pole of the Earth's great magnet.

When a hot molten rock – a magma or lava – cools and solidifies, the magnetite crystals within the cooling lava are frozen in alignment with the field direction at the time. A similar process happens when sediments lithify, or change from a wet slurry to solid rock: the solidification locks the aligned magnetic particles in place. If they are present in sufficient quantity, they give their enclosing mother rock a magnetic signal.

If the exact orientation of the rock is known, a piece of the rock that is carefully removed and taken to the lab can indicate the actual direction of the Earth's magnetic field when the rock was formed. This will show whether the rock in question was formed during a period when the north pole of the Earth coincided with the positive pole of the Earth's magnet, or vice versa.

ERRATIC SURPRISES

The magnetic reversal record discovered in the thick white limestone of Italy revealed several surprises. First and foremost was the discovery that, for more than half of the 60 million-year-long Cretaceous Period, there were no magnetic reversals at all. Until this discovery, it was assumed that the cadence of magnetic reversal was rather constant, with a shift every half-million to million years or so. But for reasons that are still unclear the mechanism creating field reversals occasionally went on holiday. From about 118 million to about 83 million years ago, the magnetic field was locked into normal polarity.

This long interval makes magneto-stratigraphic dating useless for this interval. However, it also makes the first reversal after the long interval one of the most recognizable in the entire geological column. Anywhere on Earth, if you know that you are looking at rocks formed in the Cretaceous Period, a reversed polarity signal following a long interval of normal polarity tells you precisely where you are in the geological column – at 83 million years ago.

CONTINENTAL DRIFT

The paleomagnetic sampling of the Italian Appennines was prompted by an investigation into movements of the Earth's crust – movements that, until comparatively recently, were denied by many eminent authorities as scientific heresy. What could be more stable than the surface of the Earth? Yet anyone living in earthquake country knows the fallacy of that statement, as well as why there are earthquakes. It is because continents drift.

The concept of continental drift first came under serious consideration in the late nineteenth century, when the Austrian geologist Eduard Suess suggested that Africa, Madagascar and India were once all joined as a single landmass which he named Gondwanaland. Suess based this proposal on the similarity in rock types to be found in all three areas. Suess was no charlatan or crackpot, and soon a few other geologists began considering the possibility that a 'supercontinent' existed in the southern hemisphere during the late Paleozoic and early Mesozoic Eras.

CONVERGENCE

TRANSFORM

DIVERGENCE

ABOVE **Tectonic plates either crash together (convergence), slide horizontally past each other (transform) or pull apart (divergence).**

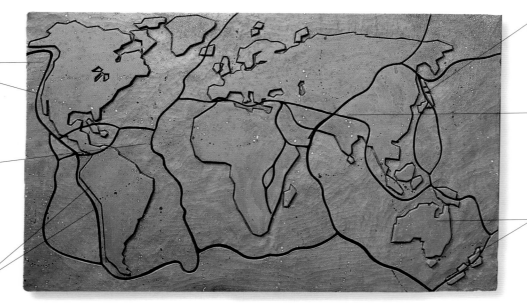

The bed of the Pacific is sliding north under the coast of Alaska, carrying western California along for the ride

A spreading seam down the middle of the Atlantic Ocean is driving America and the Old World further apart every year

As it ploughs under western South America, the Pacific floor dives into a deep ocean trench and rucks up the edge of the continent into the fiery, majestic Andes

Created by the volcanoes that form the Pacific 'ring of fire', Japan is constantly rocked by earthquakes as the Pacific pushes at its boundaries

The collision of India and Asia has created the vast crumple zone of the Himalayas and Tibet, where the crust is thicker than anywhere else on Earth

The whole of Australasia has drifted around the globe, carrying the strange animals that first evolved when Australia was attached to South America and Antarctica

ABOVE **The crust of the Earth is a complex arrangement of curved plates, like the panels of a stitched football. Some plates are moving apart, and new crust is forming in the gaps. Other plates are grinding together, pushing up great mountain ranges punctuated by volcanoes.**

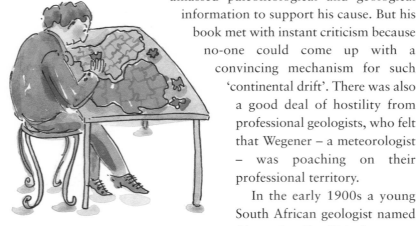

ABOVE **By examining the coastlines of western Africa and eastern South America – fitting them together like a jigsaw puzzle – Wegener concluded that they could have been part of a supercontinent.**

AN OUTRAGEOUS HYPOTHESIS

The various threads supporting the concept of an ancient, southern supercontinent were brought together in a remarkable book published by German meteorologist Alfred Wegener in 1912. Wegener was convinced that the great similarity in coastlines between western Africa and eastern South America went far beyond coincidence. He amassed paleontological and geological information to support his cause. But his book met with instant criticism because no-one could come up with a convincing mechanism for such 'continental drift'. There was also a good deal of hostility from professional geologists, who felt that Wegener – a meteorologist – was poaching on their professional territory.

In the early 1900s a young South African geologist named Alexander Du Toit began to criss-cross South Africa, spending 20 years examining rock structures, mapping huge expanses of territory and, in the process, filing vast amounts of information away in his encyclopedic memory. Du Toit soon realized that Wegener's outrageous hypothesis explained many of the geological features of southern Africa, and in 1921 he published his first paper on the possibility of 'continental sliding'.

Geologists had long been puzzled about the origin of the mountains rimming the South African coastline. The answer, realized Du Toit, was that they could have been compressed by continental collision: he had a vision of southern Africa caught in a monstrous vice between South America and Antarctica. Du Toit was able to visit other southern continents, where he observed remarkably similar successions of rocks. It was not just that the rock types were the same: the most powerful argument was that they showed a similar pattern of layering, or stratigraphy.

Wegener and Du Toit turned out to be correct. But like Vincent van Gogh, who never saw his genius acknowledged, neither Wegener nor Du Toit lived long enough to see their great triumph of observation and reasoning confirmed. Wegener himself died while attempting to cross the Greenland ice sheet in 1930. The proof of wandering continents did not burst into the scientific consciousness until the early 1960s, when a slew of studies brought the theory of a static Earth tumbling down in disarray. We now know that not only the continents but the entire surface of the Earth is on the move.

DU TOIT'S EVIDENCE

Du Toit went far beyond Wegener in his understanding of Gondwanaland. Through his examination of stratigraphy in several continents he was able to reconstruct both the early merging of the various continental pieces and their climactic melding into a single supercontinent in the late Paleozoic Era. He also reconstructed their fragmentation during the succeeding Mesozoic and Cenozoic Eras. Perhaps his most telling argument was his demonstration of the remarkable similarity of late Paleozoic rock sequences on the various continental pieces. On each he saw a basal unit of glacial tillites (coarse-grained rocks formed from glacial debris), overlain by a shale containing fossils of a small aquatic reptile called *Mesosaurus*. The shales were succeeded by deltaic and river deposits, and finally Mesozoic-aged basalt. He called this the Gondwana System, known today as the Gondwana Sequence.

PLATE TECTONICS

First, studies on rock magnetism showed that either the position of the magnetic north pole had moved through time, or that the continents had. Both seemed equally impossible. But then it was demonstrated that the line of undersea mountains known as the mid-Atlantic ridge was a linear chain of active volcanoes, constantly creating new oceanic floor. Next, a programme of deep-sea drilling demonstrated that the age of the ocean floor increased with its distance from such 'spreading centres'. This discovery showed that the ocean floor was spreading, and in many cases carrying continents with it.

But where was all this new ocean floor going? Seismic studies showed that in many places oceanic crust dips down into the Earth itself, along 'subduction zones' which are associated with mountain chains and active volcanoes. So within a few years a scientific revolution had occurred. We now know that continents have indeed drifted, and drift still. They do so because they float.

All continents are masses of relatively low-density rock embedded in a mass of more dense material. Essentially they float on a thin bed of basalt which forms the ocean floors. Earth scientists like to use the analogy of an onion; the thin, dry and brittle onion skin can be thought of as ocean crust, sitting atop a concentric globe of higher-density, wetter material. The continents are like smudges of slightly different material incorporated in the onion skin.

Unlike an onion, however, the Earth has a radioactive core, and constantly generates great quantities of heat as the radioactive minerals, entombed deep in the interior, break down into their various daughter products. As this heat rises towards the surface it creates gigantic convection cells of hot, fluid rock in the mantle, the thick layer of material lying directly beneath the Earth's crust.

Like boiling jam, the viscous upper mantle rises, moves parallel to the surface of the Earth for great distances (all the while losing heat) and then, much cooled, settles back down into the depths of the Earth. These convection cells carry the thin, brittle outer layer of the Earth along with them as plates of rocky crust. Some of these plates are composed only of ocean bed, but some carry continents or smaller landmasses trapped in the moving outer skin. This process, known as plate tectonics, is one of the great unifying theories formulated by the scientific method.

BELOW **The theory of tectonics shows that the Earth's surface is not as solid as once thought. In fact it is broken into 20 or so giant slabs of rock which are continually shifting to and fro.**

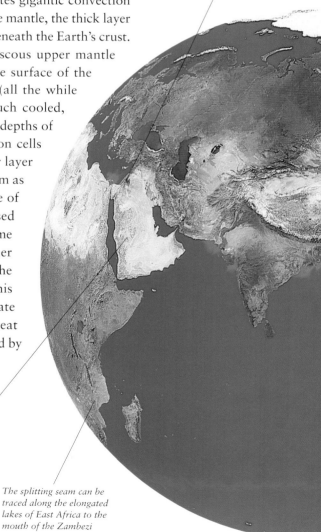

The Rift Valley extends north into the lands to the east of the Mediterranean, and at the Dead Sea great chunks of the Earth's crust have subsided into the mantle below

Arabia is swinging away from Africa on a separate plate, while molten rock wells up beneath Djibouti to fill the gap

The splitting seam can be traced along the elongated lakes of East Africa to the mouth of the Zambezi

ACCRETING CONTINENTS

The enormous plates of the Earth's crust can interact with other plates in only three ways: they can smash together, pull apart, or slide alongside each other. The first two – the divergence of plates at the spreading centres of mid-ocean ridges, and their convergence at subduction zones – were the subjects of the pioneering wave of plate tectonics research, which demonstrated the reality of continental drift. Yet the third, the sliding motion of plate margins, is equally important.

The San Andreas Fault in California is perhaps the best example of sliding motion. The San Andreas has been operating now for millions of years, and as the thin slice of California on the Pacific side of the fault scrapes northward, earthquake by earthquake, it separates rock units which were once joined together.

CONTINENTAL COLLISIONS

The collision of two continents is a slow, majestic process. Since they move at only a few centimetres per year, thousands of lifetimes must pass before any position change becomes apparent. But the converging continents move inexorably closer, and eventually they collide.

The first contact between the continental shelves does little. But as the two blocks of continental crust coalesce, enormous forces of compression act on the continental edges until the outermost regions buckle. Mountains begin to form, often punctuated by volcanoes spewing lava and ash onto the contorted mass of rock and sediment that was once a tranquil coast of wide sandy beaches.

Finally the two continents become incapable of further compression. Slowly, one continent begins to slide over the other, often doubling the thickness of crust at the edges in the process. Then, no longer able to give any more ground, they lock together.

A relatively recent and dramatic example of this awesome process is the collision of India and Asia. Forty million years ago, India was a small fugitive from the ancient Gondwana supercontinent, fleeing northward from its Southern Hemisphere origins until it collided with mainland Asia. The edge of the Indian continent rode up onto the Asian mainland to create the Himalayas, the thickest region of continental crust on Earth.

The San Andreas fault, and the trajectories of the two rock masses it separates, also provides an excellent example of how many continental margins interact. One of the great discoveries of the continental drift model is that continents not only trundle across the surface of the globe like lugubrious bumper cars, alternately smashing into each other and drifting apart, but they also grow through plate tectonic processes. Small slivers of other landmasses – such as islands or pieces of other continents – drift ashore on the edges of larger continents, meld into them through the long eons of time and inexorably enlarge them, in a process known as accretion tectonics.

Plate tectonics gave us the overall view of how the largest pieces of Earth's crust have interacted through time. Accretion tectonics tells us about continental assembly – or it should do. But unfortunately the whole subject is fraught with difficulty, particularly when it comes to understanding mountain belt formation. Since mountains are composed of rocks thrust and shattered, folded and heated, pressurized and uplifted, they are usually impervious to simple structural analysis. In short, the trauma of mountain building destroys the evidence.

When mountains are formed by only one or two processes – such as compression or extension, or even by the formation of high volcanoes – it is fairly easy to work out what has happened. But when such processes are combined with accretion things become much more complex.

Today, the most intensively researched aspects of continental drift relate to such collisions between continents and small crustal fragments. These ephemeral and active bits of land are called suspect terranes.

MOBILE FRAGMENTS

Terranes – and the suspect ones among them – can be fragments of continents composed of old sedimentary rocks, and perhaps granites and old metamorphic rocks. Others are fragments of the volcanic island arcs that form along destructive boundaries between ocean plates, and are made up of accumulated lava and sediment.

In each case the accumulated rock is buoyant, relative to the dense oceanic crust, and so 'floats' on the heavier material. Carried across the globe by the mobile plates, it sooner or later collides with some other fragment of buoyant crust, or even with a larger landmass.

ABOVE **Iceland is one of the few places where a spreading ridge is visible above sea level. Peppered with volcanoes, it is created almost entirely from basalt lava which often fractures into columns as it cools.**

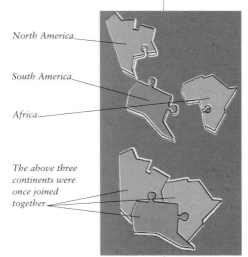

North America

South America

Africa

The above three continents were once joined together

ABOVE **The continents of North America, South America and Africa could be slotted together like the pieces of a giant jigsaw – because they were once a single piece.**

TECTONIC CHAOS

Most tectonic plates are enormous affairs, covering significant portions of the Earth's crust. The reason why there should be so few plates making up the surface of the Earth is one of the great geological mysteries. North America, for instance, although gigantic in itself, is only one piece of a plate, which also includes half the North Atlantic. The plate that makes up most of the vast Pacific Ocean is even larger. But interspersed between these behemoths are smaller plates, or microplates. When microplates and larger plates interact, quite complex geological events can occur.

Take, for example, the region around Indonesia. It rests on a microplate squeezed between three behemoths: the Indian-Australian plate, the Pacific plate and the Eurasian plate. Australia is moving northward, slowly squeezing all of the myriad islands around Indonesia into Asia in the process.

The result is geological chaos: the collision is smashing volcanic island chains, sediment-filled basins, pieces of continental margins, oceanic seamounts, and young oceanic crust into the same geological blender. Each of these units is composed of its own rock type, and as they crunch together and ooze over one another they form a tangled mess. Some pieces are raised up, others are depressed and run over; mountain ranges form, mountain ranges are destroyed. Pity the poor geologists of the future, 50 million years from now, trying to deduce the sequence of events. But just such tectonic chaos appears to have formed the West Coast of North America, 50 to 100 million years ago, and pity poor us – the geologists of today – trying to work out exactly what happened.

MAGNETIC LATITUDE MARKERS

A rock containing paleomagnetic information can reveal more than the polarity of the Earth's magnetic field when it was formed. If the rock's original orientation is known – which usually means horizontal in the case of a sedimentary rock – then the difference between this and the vertical orientation of the micro-magnets within the rock can indicate the latitude at which the rock was originally formed.

Imagine a compass turned on its side, so the needle rocks like a see-saw. If the compass needle is sufficiently well supported it will align itself with the vertical component of the Earth's magnetic field. This means that, at the North Pole, the 'north' end of the compass needle will point straight down, while at the equator the needle will lie perfectly horizontal. At all points in between, the needle will lie at an angle which increases with the latitude. This is called the dip angle.

The micro-magnets in the rock behave in the same way. By measuring their angle of dip, a researcher can deduce the latitude at which they were frozen into the rock – but only if the original orientation of the rock is known.

BELOW **Tiny magnetic particles frozen within rocks as they formed give a remarkable insight into past changes in the Earth's surface, indicating not only the direction of the North Pole at the time, but also the rock's latitude.**

WANDERING POLES?

Tracing the movements of such mobile fragments is not easy, but one of the most useful tools is derived from the study of rock magnetism. Any igneous or sedimentary rock solidifying for the first time usually contains tiny magnets which can yield information about the Earth's polarity when the rock was formed. But they can reveal more than this. They can also reveal the precise orientation of the geomagnetic field, which is itself dictated by latitude. In other words, it can show where the rock was formed.

These two bits of information became crucial lines of evidence proving continental drift. Magnetic cores from different continents seemed to indicate very different directions for the ancient magnetic poles. Successively older cores also suggested that the position of the north magnetic pole had been migrating. The answer, of course, was that the continents themselves were wandering, and paleomagnetic studies can provide detailed histories of their drift.

Such is the upside of paleomagnetic analysis. The downside is that if the rock has been reheated, it gives spurious results. Unfortunately there is never an easily observable sign that this has happened. What's more, the 'resetting' of magnetic mineral grains happens most frequently in the mountainous regions of the Earth, where regional heating and high pressure are most intense – the very regions that enclose the various slivers and shards of crust formed by the collisions of continents and terranes.

COMPLEX INTERACTIONS

The confusion created by such problems has led to a series of disputes over the formation of western North America: one of the most mysterious continental regions to have come under close study to date. From Alaska in the north to southern Mexico in the south, virtually the entire western margin of North America is mountainous. How did these mountains form? Were they the product of subduction only, where ocean crust dives under a continental margin, melting it and

creating a line of volcanoes in the process? There certainly are many volcanoes along the mountain chain, and when the movements of the oceanic crust were deciphered in the early 1960s it became clear that many of the mountains were indeed created by subduction. The high Cascade volcanoes of the northwest Pacific seaboard are an excellent example. Equally clear, however, is the fact that this destructive plate-to-plate contact – creating subduction zones – was not the only type of plate interaction going on, and that other processes of mountain building have been taking place as well. Perhaps nowhere are these better studied than in California.

California is earthquake country. Earthquakes occur when rocks move. Most of California's earthquakes can be attributed to the second type of plate-to-plate motion, where the edges of the plates scrape past each other. Known as strike slip, this type of movement is causing the westernmost portions of the Golden State to move northward. This motion will continue until the various fragments run into another part of the continent. When they do, they will have changed the shape of North America yet again, through the process of continental accretion.

SLIPPING NORTH

It now seems probable that much of Western North America was created by just such collisions with smaller landmasses, forming mountains in the process. We are witnessing the start of one of these episodes. San Francisco and the long sliver of land it sits on will, over the next few million years, leave North America to become a long borderland west of the continent, creating an inland sea in the process. Eventually it will collide with North America again, and this seems to have been the tectonic style for this whole area for several hundreds of millions of years.

If this is true, then where did the landmasses come from? Earth scientists began to wonder if much of the west coast of North America had formed far to the south, and then become accreted onto the continent. In the early 1970s, teams from Canada and the United States began sampling various rock bodies for paleomagnetism to test such ideas. This data was to became the primary weapon in a war pitting the geophysicists against many geologists, who believed that plate tectonic models requiring the transport of rock bodies over huge distances were not necessary to explain the geology of western North America.

ABOVE **At 5.12am on 18 April, 1906, a section of the San Andreas fault snapped near San Francisco, California. In Marin County, just north of the city, the rocks on the Pacific side of the fault moved over six metres to the north in 40 seconds.**

AN UNEASY ALLIANCE

There has long been an uneasy alliance between the two great pillars of Earth science: geology and geophysics. In its simplest sense, geology deals with when the observable features on and in the Earth formed; geophysics is interested mainly with the physical forces which did the forming. So geology is more concerned with time and geophysics with process. Both fields borrow from the other, and the distinction between them is anything but clear-cut. Nevertheless, while geologists and geophysicists usually work in concert, they have often been on opposing sides of fierce arguments. Two prime examples of this scientific strife afflicted the debates over the most important questions confronted by the geological sciences: the age of the Earth, and the concept of continental drift.

With regard to the age of the Earth the great Lord Kelvin, the father of geophysics, was certainly wrong (as were his followers). These errors led many scientists astray for decades. The geophysicists made a similar, monumental blunder when continental drift was first proposed in the first half of the twentieth century, for they could not conceive of a way in which such huge blocks of matter could sail across wide ocean basins. Geologists amassed tomes of evidence, yet the geophysicists held firm in their opposition to the idea and, as with their stance concerning the age of rocks, retarded scientific progress for decades.

But more recently the shoe has traded feet. Now, in many regards, it is geology which has become obstinate, and geophysics which frets over the delay. The controversy concerns the role of suspect terranes in the accretion of continents, with particular reference to the formation of western North America. The geophysicists believe they have the evidence, but the geologists remain unconvinced. An uneasy alliance indeed.

THE CASCADE CONTROVERSY

The opening shot was fired in 1971, when R. Tessier and M. Beck of Western Washington State University published paleomagnetic data from the rocks making up Mount Stuart, an extinct volcano in the North Cascades of Washington State. Far more ancient than the larger volcanoes of the Cascades, such as Mount Rainier or nearby Mount Baker (which is still an active and exceedingly dangerous volcano), Mount Stuart has been repeatedly dated, using sophisticated radiometric dating techniques, to the Middle Cretaceous of about 100 million years ago. The dispute about Mount Stuart is certainly not about its age, but where it was formed.

The pioneering work by Tessier and Beck gave a startling result: they found that Mount Stuart was composed of rocks that apparently solidified at a latitude far to the south of its present position. If the paleomagnetic results were correct, the gurgling, bubbling, molten magma which, a hundred million years ago, solidified into what is now Mount Stuart, did so somewhere off what is now southern California or Mexico. Soon after this finding, other paleomagnetic results were published which suggested that not just Mount

The thick cloud that billows from erupting volcanos is a mixture of steam, fragments of rock blasted by the explosion, and hot globs of molten lava

While the lava is still hot and molten, magnetic particles within it can turn like a compass to point north

RIGHT **A volcano carried away from the source of its lava by tectonic plate movement is like a tap with its pipes removed. The molten rock chills and hardens in the vent, and the volcano becomes extinct. Something like this must have happened to Mount Stuart as it was carried north from its original position near Mexico.**

The molten rock known as lava, pouring from the volcano, will provide important clues about Earth's history

Stuart, but other pieces of northwestern Washington and southern British Columbia had also originated in lower latitudes. If so, then each of these rock bodies must have travelled between 2,000 and 3,000 kilometres from the south. But here was more than just simple transport. All the rock bodies were also rotated by as much as 70 degrees – as if a giant hand had grabbed the land and twisted it to the right.

But was this interpretation correct? Could the various volcanic rock bodies examined by the paleomagnetists have yielded such results if they had not been transported? Yes! answered many geologists. Exactly these results would be expected if the various sampled rocks had been tilted, rather than transported.

It is much easier to tilt a rock than transport it. Tilting takes place all the time – just look at any range of mountains and you will find rocks in every conceivable orientation, including upside down. The paleomagnetic interpretations relied on an ability to recognize the original orientation of the sampled rocks. This was a key argument used by critics of paleomagnetic studies: the rocks sampled were not horizontal, and therefore the results were spurious.

Once the lava has cooled and solidified, the magnetic particles are locked in the same direction forever – only the rock can move

The crystal rocks through which a volcano erupts is considerably older than the volcanic rock created on the surface by the cooling lava

The molten magma that supplies a volcano with lava builds up in a vast chamber underground just before an eruption

So the controversy started, pitting 'drifters', or those who believed that the various pieces now making up the western Cordillera had been assembled by continental drift, against the 'fixists', who believed that no such assemblage had taken place. The fixists did not doubt the validity of continental drift – they just believed that many continental drift theories relating to the western United States and Canada were simply figments of overwrought geophysical imaginations. Then, in 1977, a study was published defining and identifying the greatest suspect terrane of all: Wrangellia.

A SUPERTERRANE

The definition of Wrangellia, named after the Wrangell mountains of Alaska, was for several reasons a watershed event in plate tectonic study. First, the study was published by three senior and highly respected geologists – David Jones, Norm Silberling and John Hillhouse. Second, until that time most terranes were assumed to be small landmasses – island-sized at best – but the Wrangellia terrane as first proposed was enormous: a strip of land extending from Vancouver Island in the south to Alaska in the north. Finally, both geological and paleomagnetic evidence was used to support the hypothesis, rather than just one or the other.

The Wrangellia terrane is composed of Triassic-age basalt which appears to have been formed deep under the sea. More than 200 million years ago an enormous outpouring of basalt erupted from great submarine fissures. This basalt continued to burp out for many millions of years, and eventually accumulated into a mass of hardened lava, many thousands of metres thick. Finally, in the late Triassic Period, the fissure closed. Limestone began to accumulate on top of the submerged lava, and the underwater environments where the limestone was being formed were colonized by early Mesozoic creatures, including corals and many species of ammonites. Eventually the whole complex was uplifted to form a long, linear landmass.

Both paleontologists and paleomagnetists found that Wrangellia was even more far-travelled than Mount Stuart. The origins of the Wrangellia basalt were somewhere in the southern hemisphere. The entire rock body had then been carried northward by continental drift, to eventually collide with the western coast of North America.

When did this titanic collision take place? In the early 1980s paleomagnetists made extensive studies of rocks laid down in the Eocene – early in the Tertiary – at one site near Mount Stuart, and at another to the east of Vancouver Island. In both regions great thicknesses of 50-million-year-old sandstone had accumulated in basins near mountainous regions. These sedimentary rocks contain the fossilized remains of palms and other tropical plants, suggesting that they too may have come from more southern regions.

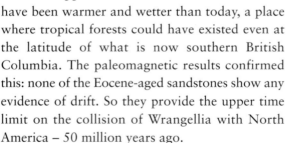

But the world in the Eocene appears to have been warmer and wetter than today, a place where tropical forests could have existed even at the latitude of what is now southern British Columbia. The paleomagnetic results confirmed this: none of the Eocene-aged sandstones show any evidence of drift. So they provide the upper time limit on the collision of Wrangellia with North America – 50 million years ago.

The idea that this long, thin terrane had smacked into North America sometime in the late Mesozoic had just filtered down into the textbooks when, in the mid 1980s, the whole concept of Wrangellia changed. Studying the rocks of British Columbia, Ted Irving, a paleomagnetist, stationed on Vancouver Island began to realize that far more than Vancouver Island and the Queen Charlotte Islands – the two prime pieces of the Wrangellia terrane in Canada – had moved up the coast. Could something much larger have moved? Soon Wrangellia received a new name: the Insular Superterrane. It now encompassed not only Vancouver and the Queen Charlottes, but all the coastal mountains of British Columbia as well. The vision now was of a huge hunk of real estate – a subcontinent as big as India, but longer and skinnier – smashing into and accreting onto the coastline. Irving coined yet another new name for this superterrane: Baja British Columbia, named after Baja California, the latitude from which he believed it had come so long ago. The entire idea became known as the Baja British Columbia Hypothesis.

As the 1980s progressed even this new landmass grew larger in the minds of the geophysicists. Now, not only the Coast Mountains and Vancouver Island, but the intermontane region – the entire suite of gigantic mountains making up most of British Columbia – was added into the mix as well. By the late 1980s, geologists from the University of Washington – Darrel Cowan and his students Paul Umhoefer, Mark Brandon and John Garver – had refined this model and put it on a far more rigorous footing by conducting motion studies of how the plate movement could have produced this event. According to all such studies, the southern limit of Baja British Columbia lies in the San Juan islands of northern Washington State, and Mount Stuart, 100 million years old, seems to have acquired its magnetic signal somewhere in the latitude of Mexico – just as Tessier and Beck suggested in the early 1970s.

The idea that much of British Columbia is really an old piece of Mexico shouldn't surprise us, knowing that the larger plates of the Earth's crust have travelled so far around the globe. Yet each such discovery is a startling reminder that plate tectonics is still a young idea, with the potential to completely upset all our preconceptions.

LEFT **Pieces of Mexico seem to have been slipping north for millions of years, welding themselves to Canada. The apparent improbability of the idea caused years of controversy among Earth scientists.**

RIGHT **The history of life on Earth is a glorious flowering of forms, evolving by natural selection into ever-increasing diversity and complexity. Yet while many of the earliest forms still flourish in all their elegant simplicity, whole categories of living things have been annihilated by mass extinctions and are now known only from their fossil remains.**

MASS EXTINCTIONS

There is perhaps only one field of Earth sciences more vibrant than plate tectonics, and that is paleontology, or the study of fossil life. The two most exciting research areas in this field deal with the Cambrian explosion of life – the period between 550 and 500 million years ago when the animals first appeared – and with the tantalizing phenomenon of mass extinctions.

Mass extinctions are global catastrophes which cause large numbers of species to be wiped off the face of the Earth. A mass extinction can occur very quickly, or it can last for between a thousand and a million years. There have been about 15 such events during the last 500 million years of Earth's history. Five of these are classified as 'major', in that they caused more than half of all species then living to become extinct.

Some 250 million years ago, the Permo/Triassic mass extinction virtually destroyed a terrestrial world ruled by the so-called mammal-like reptiles. This extinction appears to have been caused by a sudden episode of extreme greenhouse heating on a global scale. The plants and animals destroyed by this event were succeeded by an ecosystem dominated by an entirely different stock of reptiles: the dinosaurs.

The Cenozoic era saw the rise of the mammals and birds – warm-blooded animals that can survive harsh climates

The Cretaceous-Tertiary extinction that eliminated the dinosaurs is now thought to have been caused by a comet hitting the Earth

We like to think we are the pinnacle of evolution, but it is most likely that we are just the agents of the next great extinction

The Mesozoic Era was the age of reptiles such as the pterosaurs and the dinosaurs

The Permo-Triassic extinction that marks the end of the Paleozoic Era had its main impact on marine life, eliminating animals like the trilobites

The Paleozoic Era saw the development of a host of life forms that survived the great extinctions, possibly because of their relative simplicity

Sixty-five million years ago it was the dinosaur's fate to be destroyed by freak chance: the Chicxulub asteroid collision in the Yucatan peninsula region of Mexico. The dinosaurs were eventually replaced by the many species of large mammals we are familiar with today. Had that asteroid not hit, many scientists believe that there would not have been an Age of Mammals as we know it, and perhaps no humans as a consequence. So there is clearly a great deal of contingency in the history of life on this planet.

The extinction of the dinosaurs is by far the most celebrated of all mass extinctions. The proposal that it was caused by an asteroid collision was made as recently as 1980, when a group of scientists from Berkeley University in California came up with two bold hypotheses.

One of the prime movers was Walter Alvarez – the man who discovered the invaluable record of datable paleomagnetic reversals in the Italian Appennines. While he was working on the Appennine sections Alvarez collected a series of clay samples that yielded some unexpectedly high concentrations of iridium – an element that is more abundant in asteroids than it is on Earth. This led to the now-famous hypotheses that, some 65 million years ago, the Earth was struck by an extra-terrestrial object of some sort – such as an asteroid or comet – and that the environmental effects of that impact brought about the total, permanent destruction of many of the animals and plants then living.

Today these hypotheses are widely (although not universally) accepted, and the excitement and debate they generated have changed the landscape of geological and especially paleontological research. Since that time, the study of extinction has been one of the driving forces of contemporary paleontology, and has given rise to ideas that were surely unimagined in 1980. The revolution in thinking is not that asteroid or comet impacts can be the cause of mass extinctions, but that mass extinctions, through whatever cause, can occur alarmingly quickly.

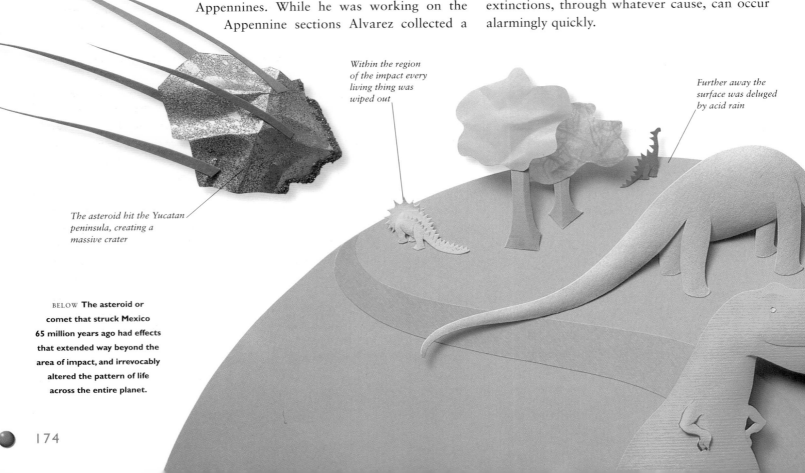

The asteroid hit the Yucatan peninsula, creating a massive crater

Within the region of the impact every living thing was wiped out

Further away the surface was deluged by acid rain

THE CRATER OF DOOM

Since 1980 the event that caused the extinction of the dinosaurs – the Cretaceous/Tertiary or K/T mass extinction – has been the subject of intensive research. It seems that there was a single comet strike, coming 1–3 million years after two rapid changes in global sea level which were themselves sandwiched around a major change in ocean water chemistry. The impact created a large crater in the Yucatan peninsula, up to 300km across. The crater, now known as Chicxulub, was subsequently buried and is no longer visible on the ground. Yet although there is still some debate about its actual size, there is no doubt that the structure is an impact crater.

The local geology and geography may have maximized the lethal effects of the strike. At the time, 65 million years ago, the area was a shallow equatorial sea with a bed of sulphur-rich evaporites (minerals deposited by evaporation, like the salt that rims salt lakes). The comet itself may also have contained sulphur. The colossal energy of the impact seems to have vaporized all this sulphur-rich material, with unbelievably dire consequences: a worldwide change in atmospheric gas content accompanied by a drop in temperature, acid rain (mainly sulphuric acid derived from the sulphur at the impact site) and global wildfires. Most scientists also agree that the thick, coarse-grained deposits found at many places along the eastern coast of Mexico were formed by impact waves. All these effects could have been lethal on a global scale, and could easily have accounted for the catastrophic disruption of the ecosystem which destroyed the dinosaurs.

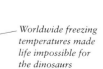

Worldwide freezing temperatures made life impossible for the dinosaurs

KILLING CLOUDS

ABOVE **The clouds of dust thrown into the atmosphere by a comet impact could have created a dark, cold world, hostile to many of the plants and animals that had evolved during an era of tropical warmth.**

The Chicxulub comet probably killed relatively little as a direct result of its impact, even though the shock of that impact must have been colossal. Its effect on the composition of the atmosphere, however, was far, far more lethal. Recent studies have shown that a huge mass of sulphur was released into the atmosphere after the impact, estimated at between 40 and 700 billion tonnes. A small proportion of this was converted into sulphuric acid which fell back to Earth as acid rain. This may have been a killing mechanism, but more disastrous to the biosphere was the probable reduction – by as much as 20 per cent – of the solar energy reaching the Earth's surface, for some 8 to 13 years. According to current models this would have been enough to produce a decade of freezing or near-freezing temperatures, afflicting a world that, at the time of impact, had been largely tropical. This prolonged 'winter' would have been quite sufficient to bring about the worldwide mass extinction.

Another recently published model suggests that greatly increased levels of atmospheric dust generated by such an impact may be lethal as well. The fine dust would cause a blackout lasting for several months. The reduction in light levels (below that necessary for photosynthesis) would be accompanied by rapid cooling of the land. The dust would also adversely affect the world's hydrological cycle. Advanced climate modelling has indicated that, following a large impact event, average global precipitation would decrease by more than 90 per cent for several months, and would take years to return to normal.

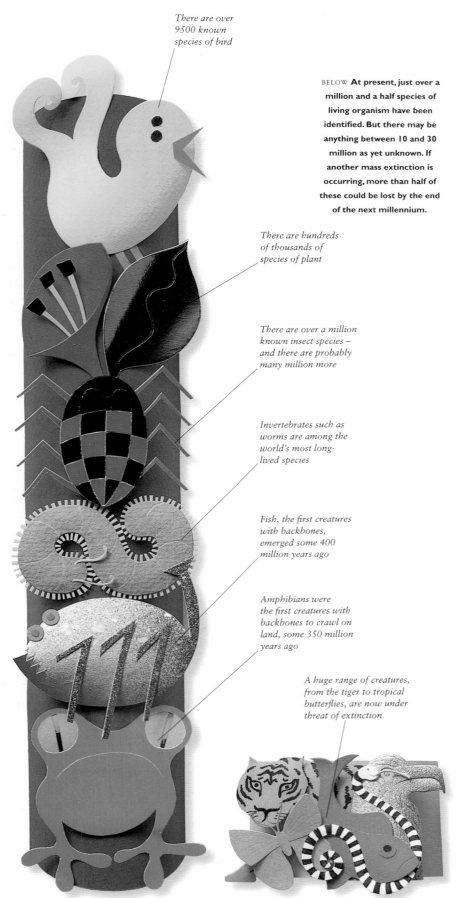

There are over 9500 known species of bird

There are hundreds of thousands of species of plant

There are over a million known insect species – and there are probably many million more

Invertebrates such as worms are among the world's most long-lived species

Fish, the first creatures with backbones, emerged some 400 million years ago

Amphibians were the first creatures with backbones to crawl on land, some 350 million years ago

A huge range of creatures, from the tiger to tropical butterflies, are now under threat of extinction

BELOW **At present, just over a million and a half species of living organism have been identified. But there may be anything between 10 and 30 million as yet unknown. If another mass extinction is occurring, more than half of these could be lost by the end of the next millennium.**

There is still much to learn about the timing, details, and mechanisms involved in the K/T catastrophe, and there are probably decades of research necessary to flesh out the picture. Yet we have learned some important lessons. We now know that impacts can cause species' extinctions, and that there is usually more than one killing mechanism going on in any extinction.

In the early days following the Alvarez hypothesis, some investigators thought that a 'general synthetic model' would emerge, linking most or even all mass extinctions to impact events. This was the thinking behind the 'Nemesis hypothesis' of astronomer Rich Muller from Berkeley. It was also the implication behind the work of David Raup and Jack Sepkoski of the University of Chicago, who suggested in 1984 that mass extinctions show a 26-million-year periodicity that could be linked to certain astronomical phenomena.

It also led to David Raup's concept that impact events would fall on a kill curve: the bigger the impact, the more sweeping the extinction. This may seem obvious, but many variables must come into play, including factors associated with the incoming body (its size, composition, angle of impact and velocity), as well as the nature of the target area. Judging by Chicxulub, the geology of the target region may have profound implications for the lethality of the strike. Moreover, not only the geology of the impact site, but also its geography may play an important part: an impact in a low-latitude site will have entirely different consequences from a similar body with similar angle and speed striking a high-latitude site, since the distribution of lethality across the globe may be produced by atmospheric circulation patterns.

Finally, the nature of both life and the atmosphere at the time of impact are important. An impact on a highly diverse world of specialists must have different effects from an impact on a low-diversity world of generalists. And an impact on a greenhouse world may have different effects than an impact on a world where the atmosphere is less rich in greenhouse gases than it is today.

ABOVE The woolly mammoth was a large, elephant-like creature that once roamed the tundra of northern Asia. A number of intact carcases have been found, frozen and preserved in the permafrost – like this baby found in Siberia in 1989. The woolly mammoth is just one of the many major species that have been lost in the last 10,000 years.

COSMIC BLAST

One potential cause of a mass extinction is an exploding star – a supernova – in the Sun's galactic neighbourhood. Two astronomers from the University of Chicago have calculated that a star going supernova within 30 light years of our Sun would release fluxes of energetic electromagnetic radiation and charged cosmic radiation that would be sufficient to destroy the Earth's ozone layer in 300 years or less. Recent research on ozone depletion in the present-day atmosphere suggests that removal of the ozone layer would prove calamitous to the biosphere, since it would expose both marine and terrestrial organisms to potentially lethal solar ultraviolet radiation. Photosynthesizing organisms such as phytoplankton and reef communities would suffer particularly badly.

By calculating the number of stars that have been within 30 light years of the Sun in the last 530 million years, and the rate at which supernova explosions occur, the astronomers concluded that there could have been one such explosion every 200 to 300 million years, with traumatic consequences for the development of life on Earth.

THE CURRENT CATASTROPHE

Could a mass extinction happen again? Abundant scientific evidence now shows that our world may have already entered another phase of mass extinction. Many prominent victims have already disappeared: over the last 10,000 years we have lost a host of larger land animals such as sabre-toothed cats, woolly rhinos, cave bears, American lions, elephant birds, moas, Irish elk, and a hundred other giant creatures that once graced this world. They were exterminated by some combination of climate change, human predation and habitat disruption caused by the emergence of agriculture.

During the coming centuries this loss of larger mammals will almost certainly continue, due to the ever-enlarging human population. Humans will undoubtedly survive, but many – perhaps most – other species now living on Earth might not. Estimates of the final extinction tally vary, but a majority of life scientists estimate that at least 25 to 50 per cent of all species – large and small – could be extinct by the end of the next millennium.

Proving this contention is exceedingly difficult. One problem, of course, is that since we have only the faintest idea about how many species there currently are on Earth, we cannot tell what percentage of total species biodiversity is actually disappearing each year. Approximately 1.6 million species have been defined to date, but most taxonomists agree that there are far more, especially among tropical insects. Peter Raven of the Missouri Botanical Garden estimates that there are a minimum of 10 million species, while in 1993 E. O. Wilson suggested that there may be as many as 30 million species.

Yet if we have such a poor handle on how many species there are, how can we arrive at a reasonable estimate about how many are going extinct? It is likely that, for every known extinction, many 'unknown' species are also eliminated, so the actual numbers involved are irrelevant. In percentage terms, we are almost certainly presiding over a biological catastrophe that is already well underway.

Certain facts are inescapable. The human population will double to more than 11 billion people by the end of the next century; humans are large animals that monopolize resources; humans cause extinctions. Certainly, a significant proportion of large mammals has already become extinct in all continents save Africa in the last 40,000 years. And certainly the rampant reduction of forest cover in the world, especially in the tropical rainforests, is leading to wholesale species extinction. So at what point, if ever, will enough species be killed off to give us the dubious distinction of participating in a recognized mass extinction event? The most extreme estimate I have heard comes from Peter Raven: in 1995 he suggested that 60 per cent of all species now on Earth will be extinct by the year 2300. As current world biodiversity seems to be far higher than at any time in the past, this will, if it comes to pass, make the current crisis the most devastating mass extinction of all time.

ABOVE **The biosphere could manage very well if the human population were kept under control.**

ABOVE **As it is, the planet is groaning under a weight of humanity that could soon become unbearable.**

THE RARE EARTH HYPOTHESIS

Today Earth scientists have a new pre-occupation – one that is shared, to some degree, by almost everyone on the planet. Are we alone in the Universe?

It is probable that, as life forms, we are not alone. Yet while life of a very simple kind, at the level of bacteria, is probably very common on other worlds (and in other galaxies), multi-cellular animal life – which we can define as motile organisms of many cells which depend on other organisms for food – is probably rare. Why? Because the diversity of complex life on Earth has evolved through an extraordinary set of physical conditions and chance events.

We now know that although other planets do exist in the Universe, few of the planetary systems discovered so far are remotely similar to our Solar System. We know as well that the fossil record of the last 3.5 billion years of Earth history shows a relatively abrupt transition from single cells to complex body plans, and that this diversification was aided by newly evolved gene systems. This sudden diversification of animals, often called the Cambrian explosion, occurred about 500 million years ago, more than three billion years after life first originated here on Earth. So evolution did not gradually create complex animals. When they finally evolved they did so quickly, in response to a set of environmental conditions that were quite different from those which allowed the evolution of life in the first place.

It seems that the evolution of complex animals requires a large number of independent and low-probability events which are different from those needed to originate and maintain microbial life. Once evolved, these complex animals are highly susceptible to mass extinctions. The maintenance of animal life on a planet depends on some critical threshold level of diversity, which allows the animals to survive the inevitable mass extinction events that will occur throughout any planet's history.

Taking all this into consideration, its seems that the evolution, and then maintenance of advanced life requires a planet of a certain type, lying in the right orbit within a solar system offering both dependable energy and protection from stray asteroids. It also depends on many other conditions that, on Earth, have arisen through luck. The chances of such lucky coincidences occurring elsewhere in the Universe seem vanishingly small.

ABOVE **One of the most exciting fields of modern earth science is the search for life on other 'earths' – yet the chances of such planets existing seem slim.**

AT A GLANCE

FOSSIL DATING

RADIOACTIVE DATING

1715

Edmund Halley suggests that the age of the Earth might be computed from the rate of accumulation of salt in the oceans.

1740

Count de Buffon experiments with the cooling rates of heated globes to show that the Earth is much, much older than Biblical estimates.

1788

James Hutton insists that the Earth is millions, not thousands of years old as had been thought.

JAMES HUTTON

1792

John Phillips divides the Earth's history, since the time when fossils became plentiful, into three ages – the Palaeozoic, Mesozoic and Cenozoic Eras – separated by mass extinctions.

1817

William Smith shows how the sequence of rocks within an outcrop can be dated from the fossils each layer contains.

1860

William Thomson (later Lord Kelvin) and Clarence King estimate the age of the Earth at around 24 million years from its rate of cooling, assuming it started as a molten mass.

1896

Antoine Becquerel discovers uranium.

1898

Marie Curie discovers the element thorium and the property of radioactivity.

1904

Ernest Rutherford shows that the heat released by the radioactivity of the Earth's uranium and other substances means that Kelvin's calculations were wrong and that the Earth is probably much, much older than 24 million years.

1907

Ernest Rutherford suggests that radioactive decay processes could be used as geological timekeepers.

1912

Alfred Wegener is the first to seriously propose the concept of continental drift.

PALEOMAGNETIC DATING

1913

Arthur Holmes pioneers the use of radioactive decay to determine the absolute age of fossils, putting real dates to geological ages for the first time.

1947

Willard Libby develops the technique of radiocarbon dating to date objects up to 40,000 years old.

1921

Alexander du Toit's work in southern Africa lends weight to Alfred Wegener's theory of continental drift.

HISTORY OF LIFE

ERNEST RUTHERFORD

1929

Motonori Matuyama discovers that the direction of the Earth's magnetic field has changed many times in the past.

1963

Vine and Matthews find proof of seafloor spreading in the magnetic reversals recorded in the ocean bed.

DIVERGENCE

1971

Tessier and Beck show that Mt. Stuart in the USA formed 3,000km to the south of where it now is.

1972

Walter Alvarez and Bill Lowrie show that there was a very marked magnetic polarity reversal at the beginning of the Cretaceous about 83 million years ago.

1977

David Jones, Norm Silberling and John Hillhouse of the US Geological Survey identify the massive ancient terrane (continental mass) of Wrangellia in northwest North America.

1985

Ted Irving shows that Wrangellia is just a small part of a much, much bigger terrane which he calls Baja British Columbia.

1988

Darrel Cowan with his students Paul Umhoefer, Mark Brandon and John Garver show that Baja British Columbia originated somewhere near the present position of Mexico.

THE EARTH

The Pulse of Life

BIOLOGY

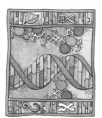

About ten billion years ago the Universe began the expansion which has created a hundred billion galaxies, each containing on average a hundred billion stars. Orbiting these stars are unimaginable numbers of planets but, as far as we have been able to determine, organic life has arisen on only one of these, our Earth.

Single-celled life forms appeared on Earth about four billion years ago, not very long after the formation of the Solar System 4.6 billion years ago. These bacteria-like organisms monopolized the planet for three billion years before multicellular animals, plants and fungi emerged in vast profusion and variety during the "Cambrian explosion" 550 million years ago. Half of the phyla – the major categories of life forms – found on Earth today originated at that time. Yet the over-whelming majority of individual species that have ever lived are now extinct, and are known only from their fossil remains in the rocks.

The study of how and why life arose on Earth and developed through time is the keystone of biology. It draws on a wide variety of special disciplines, from the field studies of the naturalist to the laboratory work of the biochemist. In particular, our present understanding of life on Earth has been shaped by two of the most important ideas underpinning modern science; the concept of the evolution of species by natural selection, first published as a theory by Charles Darwin in the mid-nineteenth century, and the molecular biology revolution, stemming from the discovery of the DNA double helix just under a century later by James Watson and Francis Crick. Darwin hypothesized that all life forms on Earth are related by common descent, and that species change by 'descent with modification'. He could not explain exactly how the mechanism of modification worked, only that it produced the natural variation fundamental to the process of evolution. The unravelling of the double helix and the decipherment of the genetic code has filled in the gaps, revealing how hereditary material mutates and passes the changes on to the next generation.

Molecular biologists have also discovered new ways of tracing the ancestry and relationships of both living and extinct species, vindicating his theory and clarifying the chronology of life. After long relying on superficial features of living organisms and a scanty fossil record, biologists can now discover exactly where we came from, and how we got here.

ABOVE **Warmed by the Sun, protected by its life-giving atmosphere and blessed with abundant water, Earth may be the only planet in the Universe with the right combination of conditions for the evolution of advanced, multi-cellular life. The story of life – how it began and how it has developed – is one of the most exciting fields in modern science.**

LEFT **Found in 1924, this two-million-year-old fossil skull of an ancestral child was the first of many such discoveries in Africa. The puzzles posed by these often fragmentary remains are now being resolved by techniques stemming from the molecular biology revolution.**

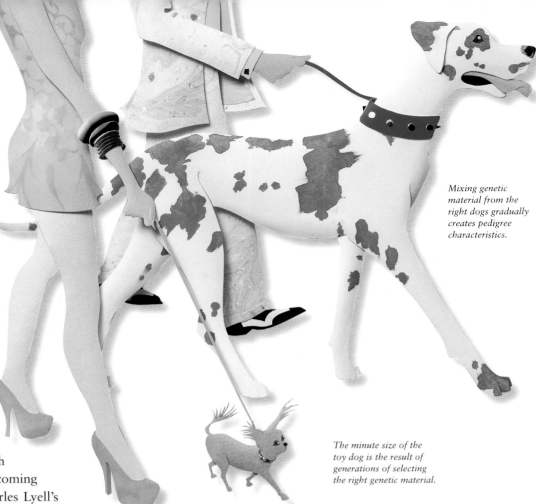

Mixing genetic material from the right dogs gradually creates pedigree characteristics.

The minute size of the toy dog is the result of generations of selecting the right genetic material.

EVOLUTION AND REVOLUTION

Biology entered the modern era in 1859 with the publication of Darwin's *On the Origin of Species*. Prior to the *Origin*, nearly everyone in the Western world believed that the Earth and all its inhabitants had been created just a few thousand years ago. They thought the geography of the Earth and the character of the individual species in it had not changed significantly since the Creation. Fossilized extinct species such as dinosaurs were 'explained' as being the races of giants alluded to in the Bible.

Darwin shattered this static, contained world-view with his vision of an ancient, dynamic Earth on which species are constantly appearing and becoming extinct. He built on the geologist Charles Lyell's view of an Earth that is always changing due to the internal forces that cause volcanoes and earthquakes. As the ages pass, the rocks of islands and continents are uplifted and eroded, revealing the fossilized remains of life forms that existed in earlier times.

Darwin proposed that all present species are derived from earlier species by the process of natural selection. Thus all species on Earth are related by descent from common ancestors. What was hardest to accept, not only by religious fundamentalists but by many of Darwin's fellow scientists, was the concept that the path of evolution has no particular direction toward 'higher forms' – humans in particular – but is determined by the particular circumstances of time and place. Darwin rejected the notion of progress.

The mechanism Darwin proposed for evolutionary change was natural selection. Just as breeders of dogs select animals with the traits they desire, to produce a new pedigree, nature favours particular characteristics under particular environmental conditions. Those individuals best suited to existing conditions are most likely to survive and pass their traits to their progeny. So evolution proceeds by descent with modification, and in time the changes give rise to new species.

Geographical separation is an important factor in generating new species, as Darwin had ample opportunity to observe during his five-year round-the-world voyage on the British Admiralty survey ship HMS *Beagle*. When plants or animals disperse to islands or are separated on the opposite sides of mountains, members of the same species may evolve and adapt in different ways because they are exposed to different pressures, and are incapable of interbreeding. Eventually their behaviour and physiology diverges to the point where they no longer recognize each other as potential breeding partners, and they become distinct species. Their common ancestor may continue to thrive in its original form or, overtaken by changing circumstances, may die out altogether, becoming another tantalizing fragment of the fossil record.

ABOVE **Genetic variation has allowed people to select and breed dogs with particular traits, and so create different breeds. The same variation provides the raw material for natural selection – the 'survival of the fittest' that has created the millions of different species. Darwin's recognition that evolution proceeded by natural selection was one of the greatest of all scientific insights.**

DARWIN AND THE BEAGLE

Charles Darwin was only 22 years old and fresh from Cambridge when he was offered the post of official naturalist on board HMS *Beagle*, a small sailing ship sent by the British Admiralty on a surveying voyage around the coasts of South America. He was little more than an amateur – he had intended to become a country parson – but his friendship with John Henslow and Adam Sedgwick, respectively Professors of botany and geology at Cambridge, had kept him abreast of the latest scientific ideas. It was Henslow who recommended Darwin for the position, giving him the opportunity to study at first-hand a wonderful variety of living animals and plants, fossils, geological formations and other natural phenomena while the surveyors did their work.

Throughout the five-year voyage Darwin collected specimens and sent them back to Henslow in Cambridge. Some of the most extraordinary and perplexing were obtained from the Galapagos islands, a volcanic archipelago lying on the Equator a thousand kilometres west of Ecuador. Darwin became intrigued by the way each island was inhabited by distinct species, each slightly different from the species on neighbouring islands. Why such variety, and how did it arise? His answer – that the various species had diversified from common ancestors through 'the preservation of favoured races in the struggle for life' – was to become the germ of his revolutionary theory of natural selection.

HEREDITY

The means by which parental traits are passed to their offspring was clearly crucial to a full understanding of this evolutionary process. When the *Origin of Species* was published in 1859, neither Darwin nor anyone else understood how heredity works. Many influential scientists – including the French biologist Jean Baptiste Lamarck and Charles Darwin's grandfather, the physician and naturalist Erasmus Darwin – assumed that characteristics acquired during the lifetime of the parents are passed on to their offspring.

If acquired characteristics are hereditary, we can account for giraffes by imagining successive generations of browsers stretching their necks to browse on higher and higher leaves. Parent giraffes who had managed to elongate their necks a little

RIGHT **For a giraffe, a long neck means more food. But how does it get the long neck in the first place?**

– just as weightlifters develop enlarged muscles – would produce young giraffes whose necks were also slightly longer than normal.

Darwin would disagree with this scenario. According to his theory, any generation of browsers would vary in neck length. If those with longer necks were able to feed better than their kin, they would tend to survive longer and produce more progeny. So natural selection would favour longer necks, and gradual elongation over successive generations would result in the creatures we now call giraffes.

Darwin's theory relies heavily on observation and common sense. Individuals clearly do vary, and this variation must affect their chances of surviving long enough to breed. But how and why do they vary, and why are these variations not mirrored precisely in their offspring?

Unknown to Darwin and the rest of the scientific world, the problem was being investigated during the 1860s by the Moravian monk Gregor Mendel. By carefully cross-breeding generations of pea plants and observing which parental traits appeared in the offspring, Mendel was working out the laws that are the cornerstone of modern genetics. But Mendel was working in obscurity, and although he published his results in 1866 they were ignored by the international scientific community (understandably, perhaps, since they appeared in the parochial *Journal of the Brno Natural History Society*). Mendel died in 1884, still largely unknown to science, and his laws were independently rediscovered by three different biologists in 1900. The widespread acceptance of these laws created a new scientific discipline, and ultimately led to the molecular biology evolution.

MENDEL'S WORK

Mendel based his laws of heredity on the patterns of inheritance he observed during eight years of experiments with pea plants. He selected seven traits – including flower position, pod colour, the shape of the ripe peas and the height of the plants – and kept meticulous records of the expression of these traits over many generations of careful cross-breeding. He then used this data to calculate the ratios in which the various traits appeared, and proposed a theory to explain them.

Almost immediately his results disproved the assumption of 'blending inheritance' which was accepted by most naturalists, including Darwin. When Mendel selected tall pea plants and short pea plants that were known to breed pure (producing offspring like themselves), and cross-bred them to create a generation of first filial, or F1 hybrids. They were all tall. There were no short plants, and certainly none of intermediate height, as you would expect if the parental traits were blended in the offspring. Yet if cross-breeding tall and short plants produced only tall plants, what happened to the trait for shortness?

The answer appeared in the F2 generation, which Mendel created by cross-breeding his tall F1 hybrids. Roughly three-quarters of these were tall, but one-quarter were short. The trait for shortness had mysteriously reappeared.

Mendel explained this phenomenon in terms of inherited 'particles' or genes that determined each trait. Every character – height, flower position and so on – was controlled by a pair of genes, inherited from the two parent plants. If each gene in the pair carried the same inherited trait for, say, shortness, then the resulting plant would be short. But if the two genes carried different inherited traits, then one would dominate the other.

Mendel's data convinced him that certain traits were always dominant, and tallness in pea plants was one such trait. So if a hybrid inherited a 'short' gene and a 'tall' gene, or t-T, the tall gene T would control the height of the mature plant. Assuming that the inheritance process was statistically random, one hybrid in four would inherit T-T and be tall, one would inherit t-t and be short, and the other two would inherit t-T and T-t and – because T is dominant – would also be tall. This explained why three-quarters of his F2 hybrids were tall and one-quarter were short.

This ratio of one to three is a fundamental law of heredity. When a number of different traits are considered the patterns become more complex, as Mendel himself discovered, but his basic concept of 'dominant' and 'recessive' characters, and the patterns of inheritance they create, remains fundamental to the science of genetics.

VITAL MOLECULES

Mendel and his followers worked out the statistics of heredity, and Mendel himself proposed the existence of the inherited 'particles' which we now call genes, yet the true nature of the hereditary material that transmits traits in plants and animals was a complete mystery to them. Nowadays we know that the molecular structure of matter is basic to gene structure and function. But although the concept of molecules had been proposed, it was far from being an accepted principle in the latter half of the nineteenth century, when Darwin and Mendel were doing their revolutionary studies. Accustomed as we are nowadays to thinking in terms of the molecular composition of both non-living and living materials, it is not easy to put ourselves into the state of mind of scientists of that era.

Many prominent chemists and physicists simply did not believe that molecules were real things. Furthermore, there was a widespread conviction that the constituents of living and non-living matter were fundamentally different, and that living tissue was endowed with some 'vital force': an ingredient missing in inorganic substances. This 'mechanism versus vitalism' debate could not be decided until the fundamental chemistry of life was worked out, for in the event the 'vital force' turned out to be a chemical phenomenon, arising from the way certain atoms – which also occur in non-living things – are put together to create special organic molecules such as proteins, amino acids and nucleic acids.

Before this synthesis could be achieved, it had to be established that atoms and molecules really existed, and were not just book-keeping devices to explain why certain materials combined in fixed ratios in chemical experiments.

For example, it had been shown that the gaseous elements hydrogen and oxygen always combined in a fixed ratio – two volumes of hydrogen to one volume of oxygen – to make water. Italian physicist Amadeo Avogadro postulated that equal volumes of each gas contained an equal number of molecules, but that number was not known.

MENDEL'S PEA PLANTS

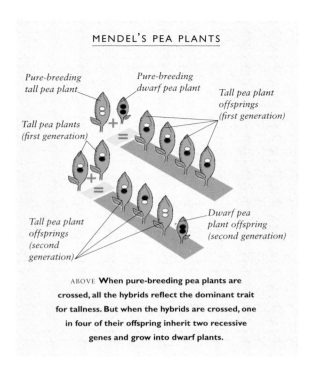

Pure-breeding tall pea plant

Pure-breeding dwarf pea plant

Tall pea plant offsprings (first generation)

Tall pea plants (first generation)

Tall pea plant offsprings (second generation)

Dwarf pea plant offspring (second generation)

ABOVE **When pure-breeding pea plants are crossed, all the hybrids reflect the dominant trait for tallness. But when the hybrids are crossed, one in four of their offspring inherit two recessive genes and grow into dwarf plants.**

In 1905 the young Albert Einstein demonstrated that molecules were real by calculating Avogadro's number. He analyzed the mysterious phenomenon of Brownian motion, named for Robert Brown, an English botanist who was a generation older than Darwin. Brown had noticed that, when viewed through a microscope, tiny particles such as pollen or soot suspended in water bounce around in random directions. Where did these particles get their energy? Scientists had debated this question for nearly a century. Some thought it came from collision between the particles and the molecules in the solution, but they couldn't prove it.

Einstein found a way of proving it. He created a formula for calculating Avogadro's number, using measurable quantities such as the size and bounce of the particles, and the temperature and viscosity of the fluid. Suddenly, it was possible to count the number of molecules in a solution, and because they were countable, molecules were transformed from rather intangible, hypothetical entities to indisputably real things.

In the mid-1920s quantum mechanics revealed the nature of the chemical bond and united physics and chemistry. Following this unification, the methods employed in chemistry were increasingly applied to the study of living organisms, and it emerged that plants and animals were made up mostly of complex protein molecules built up from

different arrangements of smaller molecules called amino acids. This implied that these two major kingdoms of living things must have had a common origin – and that their differences stemmed from some mechanism that created different types of proteins.

CHROMOSOMES AND GENES

Genetic research proceeded rapidly in the early decades of the twentieth century. Microscopic examination of plant and animal cells revealed chromosomes: thread-like structures in the nucleus of every cell (or in the cytoplasm of a bacterium, which has no nucleus). These can be seen to replicate during cell division in a process called mitosis. It seemed possible that these replicating structures carried information vital to the function of the cell.

Some organisms were found to possess particularly large chromosomes, out of all proportion to their body size. The tiny fruit fly *Drosophila* was one such, with giant chromosomes that could be inspected in some detail. Sure enough, the hereditary 'particles' that Mendel had hypothesized – what we now call genes – could be seen strung out on the chromosomes, 'like beads on a wire' according to one observer.

The fruit fly had a further advantage that it had a short life and bred rapidly, producing many generations within a short space of time. This enabled researchers to compare the chromosomes of succeeding generations, and they found that chemical alterations, or mutations, in their genes could occur spontaneously or be induced by radiation. These mutations were manifested as physical abnormalities, such as white eyes rather than the normal red, or four wings rather than the normal two. So it was established beyond reasonable doubt that the physical form of an organism was dictated by the arrangement and combination of genes on its chromosomes.

GREGOR MENDEL
1822–84
A farmer's son, Mendel carried out his breeding experiments in the kitchen garden of his monastery in Brno, Moravia. He was eventually elected abbot, but his interest in biology made him a suspect figure in the eyes of the church.

BELOW **The fruit fly's remarkably large chromosomes and superfast breeding abilities make it ideal for genetic research.**

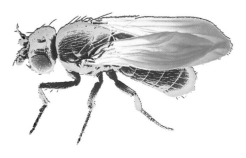

THE CHEMICAL LANGUAGE OF HEREDITY

The major problem in genetics that still had to be solved was the chemical nature of the genetic material. In the 1920s it was established that chromosomes are composed of proteins and a substance called deoxyribonucleic acid, or DNA for short. Therefore it seemed evident that genes, which are components of chromosomes, are also composed of DNA or proteins or both.

Most biochemists and geneticists favoured proteins. These are the main component of living tissues, and are made up of combinations of 20 different amino acids. If these amino acids are considered as the alphabet of a chemical language, proteins clearly have the capability of storing and conveying an almost infinite amount of biological information.

DNA, by comparison, is a complex structure built around sequences of only four chemical

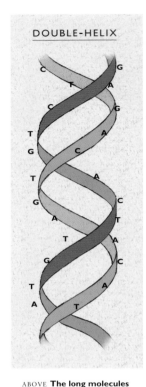

DOUBLE-HELIX

ABOVE **The long molecules of DNA are built up from paired spiralling strands of sugar-phosphate bases. Each base in a spiral strand is part of a nucleotide, which links to a complementary nucleotide in the other spiral to create the double-helix structure. Certain bases always lie opposite each other: Adenine (A) is paired with thymine (T); thymine with adenine; guanine (G) with cytosine (C); and cytosine with guanine.**

compounds called bases. At first the various combinations of this four-part sequence were thought inadequate to serve as a chemical alphabet of life. However, some elegant experiments with *Pneumococcus*, the bacterium that causes pneumonia, showed that DNA, and not protein, accounted for the difference between the infectious and non-infectious varieties. So the information determining the form and function of *Pneumococcus* was carried in its DNA. But what was DNA, and how did it work?

THE DOUBLE HELIX

In 1953 James Watson and Francis Crick figured out the structure of DNA. The new era of molecular biology and molecular genetics can be dated from that year. Before 1953 it was known that DNA, or deoxyribonucleic acid, is made up of units called deoxyribonucleotides, each with three components: a sugar containing five carbon atoms called deoxyribose, a phosphate group and a nitrogenous base. There are four different deoxyribonucleotides, or nucleotides for short, each possessing identical sugar and phosphate groups but different bases. These are adenine, guanine, cytosine and thymine. What was not known was how these components were arranged in the genes.

Watson and Crick deduced that DNA in the cell consists of two long molecular strands coiled about each other to form a double helix. The two helical strands are connected, like the steps of a spiral staircase, by weak links called hydrogen bonds between particular pairs of bases. Adenine on one strand is always matched to thymine on the other, thymine to adenine, guanine to cytosine, and cytosine to guanine. This complementary system ensures that the sequence of bases on one strand determines the base sequence on the other.

The linear sequence of bases in each molecule of DNA act as a code for the genetic message, and since the sequence on one strand of the helix determines the sequence on the other, it provides a mechanism for replicating the code by building up new strands. Special protein enzymes called

THE GENETIC CODE

The only way the four letters of the RNA alphabet can be translated into the 20 letters of the amino acid alphabet is by using combinations of the shorter alphabet. So different combinations of the four RNA letters code for different amino acids. But how many letters are needed for each coding sequence, or 'codon'? The RNA molecule is just a string of letters in apparently random order, so there are no obvious groupings. The size of the codon had to worked out by simple logic.

If each codon consists of a two-letter combination such as AA, AG and GU, then there are $4 \times 4 = 16$ possibilities. Since the code has to specify 20 amino acids this is not quite enough. But if each codon consists of three-letter combinations such as AAA, ACG and GUA this gives $4 \times 4 \times 4 = 64$ possibilities; more than enough. Therefore the genetic code is 'redundant', with an average of about three ($64 \div 20$) three-letter codons for each amino acid.

polymerases unzip the strands of the double helix into single strands. The single strands hook up with spare nucleotides floating in the liquid surrounding the molecule, with each base in the unzipped strand latching onto its complementary base. The sugar and phosphate groups of the newly-acquired nucleotides then join together into helical strands, so two identical double-helix molecules are formed, exactly like the original.

The elucidation of DNA structure provided the basis for understanding one of the key characteristics of life: the 'vital force' that enables biological molecules, genes, cells and organisms to replicate themselves. But in itself it did not explain the variation that provides the raw material for natural selection.

DNA INTO PROTEIN

The DNA in the genes is a set of instructions that directs the cell, which is like a small factory, to make the thousands of proteins needed for body structure and function. Somehow the four-letter alphabet of DNA is translated into a 20-letter alphabet: the 20 amino acids that, in different sequences, make up proteins.

In the process of protein synthesis, the DNA message is first transcribed into an RNA message. RNA is very similar to DNA, but its sugar component is ribose rather than deoxyribose, and its four bases are adenine, cytosine, guanine and uracil, which replaces thymine and forms a base pair with the adenine of DNA.

Within the cell nucleus the DNA helix unzips as if to replicate itself, but instead it gathers RNA nucleotides to form a single strand of 'messenger' RNA or mRNA. The mRNA strand detaches itself and carries its own version of the four-letter DNA message out of the nucleus into the surrounding cytoplasm, which contains free amino acids. While the DNA helix zips itself together again, the mRNA message is used by cellular organelles called ribosomes to build the long chains of amino acids that form proteins. Exactly how the four-letter alphabets of DNA and mRNA are translated into the 20-letter amino acid alphabet of proteins was the next mystery to be solved after the structure of DNA itself. It was first worked out in 1960 by Marshall Nirenberg.

The consequences of this work have been incalculable. Knowledge of the genetic code makes it possible to pinpoint the underlying basis of genetic disorders such as sickle cell disease and cystic fibrosis, which are caused by specific alterations in the DNA sequences of certain genes. Different DNA sequences also account for the variation between individuals of the same species, and for the differences between species. DNA analysis has enabled biologists to quantify these differences, instead of relying on the sometimes superficial evidence of form and function.

The discovery of the DNA double helix united biology with chemistry. Beginning with the great illumination brought about by the discovery of DNA structure and replication, biological understanding has rapidly widened and deepened.

LEFT **Every individual has its own distinguishing characteristics which mark it out from others of the same species. These variations are created by different DNA sequences – some inherited, some spontaneous. By analyzing these sequences biologists are now able to track the process of evolution.**

BELOW **The DNA molecule is like two molecules zipped together. When cells divide, the DNA unzips and the two sides come apart. Each side then attracts spare nucleotides and builds up a replica of its former partner to create two identical DNA molecules.**

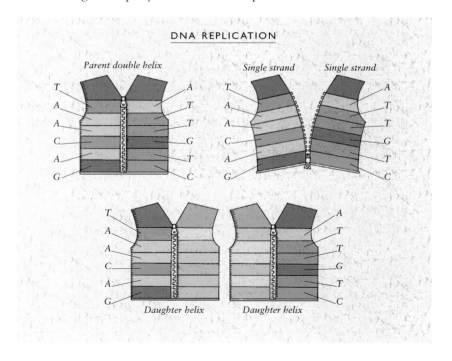

DNA REPLICATION

Parent double helix

Single strand   Single strand

T — — A
A — — T
A — — T
C — — G
A — — T
G — — C

T A
A T
A T
C G
A T
G C

Daughter helix   Daughter helix

T — A
A — T
A — T
C — G
A — T
G — C

BERRY-EATING BIRD

SEED-EATING BIRD

VARIATION

Darwin recognized that variation provides the foundation for evolutionary change. Each individual in a sexually reproducing population is unique, different in some respects from all others. Under existing conditions, some variants are more successful than others when it comes to surviving and reproducing. So nature 'selects' these variants and eliminates others. Only these selected individuals interbreed, and their traits are therefore enhanced in succeeding generations. Darwin hypothesized that this process could have produced, over long ages of time, the variety of species that occur on Earth today.

Darwin was fascinated by the enigma of variation. He devoted an entire chapter of the *Origin of Species* to the 'laws of variation', but the origin of variation continued to puzzle him all his life. Unaware of Mendel's work, Darwin believed like most of his contemporaries in 'blending inheritance' – a balanced mixture of the traits of both parents in the offspring. Mendel was the first to show that inherited traits such as colour, size and shape are controlled by discrete factors rather than being a blend of both parents. At its simplest, this is expressed in gender: the offspring of male and female parents is either male or female, and not a blend of the two.

INSECT-EATING BIRD

In fact sexual reproduction turned out to play a very important role in maintaining variation, via a shuffling of maternal and paternal genes known as recombination. This shuffling and recombination of genes in each generation maintains the variation observed in nature. It guarantees that offspring will never be identical clones of their parents and that, except for identical twins, siblings will always be genetically different from each other. But this is not the only source of variation, for inherited genetic material is also subject to the coding 'errors' known as mutations.

ABOVE AND RIGHT **The variety of different bill shapes among finches equip the various species to exploit different types of food. Darwin recognized that small natural variations in bill form could make the basis of new species, but he did not understand the genetic basis of such variation.**

MUTATIONS: THE CLASSICAL VIEW

The genetic knowledge acquired in the twentieth century explained the connection between genes and the outward expression of traits. But until the genetic code was deciphered and codon redundancy recognized, the amount of variation carried in the DNA could not be fully evaluated. That potential turned out to be far greater than was appreciated during the classic phase of genetic research prior to 1953.

Thomas Hunt Morgan – the geneticist who conducted the first breeding experiments using the fruit fly *Drosophila* – made a considerable contribution to our understanding of gene positions on chromosomes and the effects of gene changes.

NUT-EATING BIRD

He began doing research on fruit flies in 1909. One year and a million flies later he noticed the sudden appearance of a fly with white eyes instead of the normal red eyes. He cross-bred this lone male with its red-eyed sisters. In the first generation all the progeny had red eyes, so white eyes were shown to be 'recessive'. Out of 3470 progeny from subsequent inbred generations, there were 782 white-eyed males but not a single white-eyed female. So white eyes were demonstrated to be sex-linked.

In 1905 it had been learned that the X and Y chromosomes determine sex. All the ova but only half the sperm have an X chromosome. Morgan realized that, in fruit flies, eye colour is carried on the X chromosome, and an occasional mutant X carries a gene for white eyes. Since females have two X's, and red is dominant, females almost never have white eyes. Males have one X and one Y, so males with the mutant X have white eyes.

CACTUS-EATING BIRD

White eyes were not the only mutations. Another mutant fly appeared with a black body and tiny vestigial remnants of wings. The offspring of this mutant bred with a normal

GENETIC SHUFFLING

There are two types of cell division: mitosis and meiosis. Mitosis results in two cells identical to the original, and is the basic mechanism of growth and repair. Meiosis, which occurs in the cells of the reproductive organs, results in four cells which each have half the chromosome complement of the original. Instead of having two sets of chromosomes arranged in pairs, with one of each pair inherited from the mother and one from the father, the sex cell (gamete) has only a single set.

During meiosis, segments of each chromosome pair in the original cell are exchanged in a process known as 'crossing over', which shuffles the maternal and paternal genes. So the chromosome inherited from the mother acquires some genetic material from the chromosome inherited from the father, and vice versa. These 'recombined' chromosomes then give rise to the gametes – the ova and sperms. The process means that the gametes are genetically different from the body cells, and that no two gametes are exactly alike.

When the male and female gametes combine at fertilization, the full complement of chromosomes is restored. But the shuffling of the genetic material means that the chromosomes in the fertilized cell – the zygote – are not the same as those of the parents, and since one of each chromosome pair is inherited from a different parent this creates even more genetic possibilities. The potential for variation is immense. Meiosis is therefore one of the main sources of genetic variation, and one of the driving forces of evolution. The process of meiosis during sexual reproduction means that new genetic combinations are introduced with every single generation, allowing the rapid development of new characteristics.

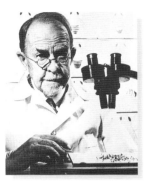

THOMAS HUNT MORGAN
1866–1945

A pioneer geneticist, Morgan started his career believing that Mendel's laws were wrong. He started breeding fruit flies to confirm his belief, but his results proved Mendel was right all along. However, they also demonstrated that inherited traits were carried on chromosomes – a discovery that earned Morgan a Nobel Prize.

RIGHT **The inheritance of gender is a simple example of the way heredity works. The female parent has two chromosomes, while the male has an X and a Y. Each parent contributes a chromosome apiece to each of their offspring, who inherits either XX or XY. This means that the offspring is either female or male, and not a blend of the two.**

fly produced two kinds: half were black-vestigial and half were normal. None were black with normal wings, or the normal grey colour with tiny wings. So Morgan deduced that the genes for body colour and wing size were linked closely together on the same chromosome, and that during the process of crossing over these genes tended to stay together. The farther apart two genes are, the more likely they are to sort separately during crossing over and not be linked.

Waiting for spontaneous mutations to show up is a slow and tedious process, even in rapidly reproducing fruit flies, and even slower in mammals. In the 1920s Herman Muller exposed fruit flies to X-rays and found that the rate of mutations was 100 times the spontaneous rate. The higher the radiation dose, the greater the mutation rate. Also, the earlier after irradiation the rate was determined, the higher it would be.

It seemed that most mutations were harmful to the organism, and only the occasional mutation was beneficial. Such beneficial mutations could lead to evolutionary change through natural selection, but the harmful mutations tended to die out for the same reason.

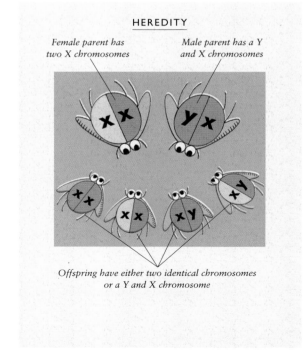

HEREDITY

Female parent has two X chromosomes   *Male parent has a Y and X chromosomes*

Offspring have either two identical chromosomes or a Y and X chromosome

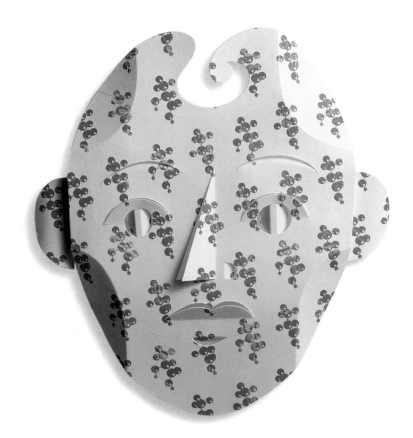

NEUTRAL MUTATIONS

In the 1960s studies of the genetic code led geneticist Motoo Kimura to propose a radical and highly controversial new view of mutation: the idea that the vast majority of mutations have little or no effect on the organism. In classical genetics, a mutation was the appearance of a new trait, such as white eyes instead of red. On the molecular level, according to the new view, mutations such as base substitutions or deletions are taking place all the time, but the vast majority are 'neutral' mutations that have little or no effect on the organism.

There are several kinds of neutral mutations. First, there are mutations in the non-coding DNA which are not translated into proteins and so have no effect on the host organism. In eukaryotes, 95 to 99 per cent of the genome consists of non-coding DNA that is not translated into proteins.

Second, there are 'synonymous' mutations in the genes, such as a mutation from the codon ATT to ATC: both are translated into the amino acid valine, so the mutation has no chemical impact.

Third, a mutation that changes an amino acid may have little or no effect on protein function.

Fibrinopeptides, for example, are components of blood clotting that simply act as spacers to keep sticky molecular surfaces apart. They are cut out and recycled when clots form. Almost any amino acid will serve their spacer function equally well, so the fibrinopeptide genes can accept more mutations than most other genes do.

So instead of the fixed genetic make-up visualized by the classical geneticists, the genomes of organisms are beehives of mutational activity – although most of this activity goes on quietly inside the cell and is not expressed.

How does this new concept of variation based on neutral mutations contribute to modern biology? It has provided a way of establishing genetic relationships which is quite independent of morphology – the outward form of the organism. So biologists can demonstrate in fine detail Darwin's 'descent with modification'.

MOLECULAR MUTATIONS

The genetic code consists of a series of three-letter 'codons', each of which translates into a particular amino acid used to assemble a protein. If there is an error in the transcription process so one of the codons contains the wrong letters, the wrong amino acid may be built into the protein. This molecular mutation may then be expressed as a physical mutation, such as those that appear among laboratory fruit flies, or it may have no effect at all, in which case it is called a neutral mutation.

Some molecular mutations are simple substitutions, so they change only one codon. But if a letter is missed out altogether it has a knock-on effect on all the ensuing codons in the row. It changes them all, with potentially far more drastic effects. The addition of a letter creates a similar situation. Such mutations can actually destroy the function of a protein altogether, and may even affect the viability of the whole organism.

JUNK DNA

In bacteria all the **DNA** in the genome acts as a code for the manufacture of protein. But in archaea and eukaryotes – which include all multi-cellular organisms – 95 to 99 per cent of the **DNA** in the genome is non-coding, and has no obvious function. This is sometimes called 'selfish', 'junk' or 'parasitic' **DNA** because it uses energy and materials in replicating itself from generation to generation without apparently contributing anything to its host.

The lengths of junk **DNA** that intervene between the coding **DNA** in the molecular strands are called introns, while the coding segments that are expressed as proteins are called exons. The introns are important to molecular biologists because they carry genetic information that is not subject to natural selection. A mutation on an exon can have such deleterious effects on the organism that it dies out, but a mutation on an intron cannot be selected against because it is not expressed. It persists for generation upon generation, and the accumulation of such neutral mutations acts as a record of the organism's genetic history.

BELOW **Aristotle was one of the first to discern some kind of order in the apparent chaos of nature – an order that Linnaeus classified, Darwin explained, and modern molecular studies are beginning to clarify in every detail.**

CARL LINNAEUS
1707–78
Brought up in poverty in Sweden, Linnaeus had an instinct for economy that was reflected in his concise, methodical approach to science. His binomial system for naming organisms had its flaws, but its ruthless logic and adaptability ensured its survival.

ORDER IN DIVERSITY

Ever since the time of Aristotle (384–322 BC) natural philosophers have tried to find some kind of order in the diversity of nature, and devoted much of their time and energy to compiling classifications of living organisms. These classifications relied mainly on comparative anatomy, and their underlying purpose changed greatly from one historical period to another.

Aristotle wanted to understand the harmony of nature, the *scala naturae*. The motive of the master classifier Carl Linnaeus (1707–1778) was to be comprehensive. Known as the 'father of taxonomy', Linnaeus invented the binomial classification of genus and species that we still use today. An animal such as the grey wolf has a two-part scientific name *Canis lupus*. The *lupus* part is specific to that species, but the *Canis* part refers to the genus: a group of dog-like creatures that share certain characteristics with the grey wolf. The genus *Canis*, for example, includes the coyote *Canis latrans* and the golden jackal *Canis aureus*.

The genus *Canis* is just one of several genera in the family Canidae, which also includes foxes such as the red fox *Vulpes vulpes* and the Arctic fox

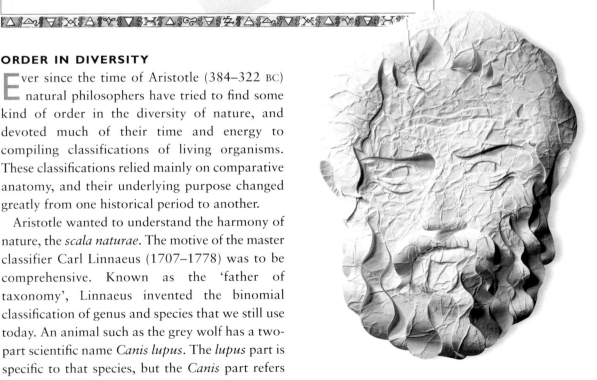

Alopex lagopus. The Canidae is one of eight families in the order Carnivora – which also includes the cats, weasels, bears, raccoons, civets, hyenas and mongooses. This order is part of the class Mammalia, which belongs to the phylum Chordata, which is part of the kingdom Animalia.

THE ORIGINS OF LIFE

Exactly how life got started is not yet known, but we can make educated guesses. Before we knew that DNA, and not protein, is the basic stuff of life, Stanley Miller at the University of California did some experiments putting high-voltage lightning-like discharges through gaseous mixtures of steam, methane and ammonia – his version of the primeval atmosphere. He succeeded in creating some of the simpler amino acids, which are the building blocks of proteins. No one has yet managed to fabricate DNA that way, but since life on Earth today requires a combination of DNA and protein enzymes for replication, the production of amino acids seems to fit one piece into the puzzle of life's origin.

By the time the Earth cooled enough for the first rocks to form, about 3.8 billion years ago, life in the shape of one-celled organisms had already arisen. Microscopic fossils embedded in ancient rocks of Australia and South Africa look much like modern bacteria. Fossil stromatolites more than three billion years old – large mats formed by colonial cyanobacteria – look very much like those that thrive in the intertidal zone of Shark Bay of Western Australia today.

Chemical evidence shows that life was abundant in the early oceans. Carbon, the element that defines organic life on this planet, is present in two stable forms or isotopes, carbon-12 and carbon-13; these have identical chemical behaviour but slightly different weights. When carbon is taken up by living cells, a little more of the lighter isotope is incorporated, with the result that organic carbon can be discriminated from inorganic carbon by being relatively poor in carbon-13. The earliest rocks are sufficiently depleted in carbon-13 to testify that life got off to a fast start in the Hadean Era, from 4.6 to 3.8 billion years ago, when much of Earth's crust was still molten and the atmosphere mostly hot steam.

ABOVE The early classifiers of the natural world had no concept of relationship through shared ancestry. They simply grouped animals and plants into genera and families on the basis of similarity. But after Darwin, these groupings acquired a new significance as parts of a spreading tree of life, all stemming from a single trunk of common descent.

The whole system was originally contrived as a means of classifying every known organism, and the place of each organism in the scheme was sometimes determined in somewhat arbitrary fashion. Physical resemblances played a large part, which is why the wolf and coyote are placed in the same genus. In fact this is probably a valid reflection of their real relationship, because they almost certainly shared a recent ancestor. But the Arctic fox and red fox were placed in different genera because they look so different, and molecular studies now show that they, too, shared a recent ancestor. So the Arctic fox should perhaps be renamed *Vulpes lagopus*. Such taxonomic reallocations are becoming increasingly frequent as molecular studies reveal the true form of the Tree of Life. Linnaeus was obsessed with including every known organism in his scheme, although of course only a small fraction of the millions of species known today (a total that is constantly expanding as 'new' species are discovered) were known two centuries ago. For the natural philosophers of the nineteenth century, who came after Linnaeus, classification was an effort to understand the plan of creation of the Great Designer of this world.

Darwin's idea of common descent replaced the metaphysical purpose of classification with a scientific one. All animals were related because they all ultimately had the same ancestors. Like members of a large family, some species were more closely related than others. The challenge of classification was to construct historically accurate family trees, showing how each species related to its contemporaries and its ancestors.

THE PULSE OF LIFE: BIOLOGY

FAMILY TREES BASED ON COMMON DESCENT

Before Darwin there was no concept of a family tree, because the different species were thought to have been divinely created all at once. Therefore they could not be related by shared parentage and descent. Animals that look alike, such as the horse, ass and zebra, were put into the same genus *Equus*, but this did not imply a genetic relationship. It was classification for the sake of convenience and nomenclature.

Once Darwin had proposed that all species stemmed from one or a very few common ancestors, it followed that they were all related in varying degrees to each other: the closer to the common ancestor, the closer the relationship. From the evolutionary point of view, the discovery of this tree of life and the degree of the relationships between species became one of biology's most important goals. Darwin drew such a tree in one of the red-bound notebooks he kept from an early age. It shows an ancestral trunk from which branches diverge, first becoming variations of the parent species and then becoming new species.

Darwin's younger contemporary Ernst Haeckel (1834–1919) proceeded to draw a literal tree of life (he was a talented artist), in which he connected all known groups of plants and animals on the basis of his conception of their common descent. Haeckel's tree still regularly appears in textbooks and is similar in broad outline to trees based on current data.

Haeckel's trees, and the family trees drawn by those who followed him, were based largely on comparative anatomy and embryology. Prior to the decipherment of the genetic code, all classifications of animals, plants, and micro-organisms necessarily relied on morphology (shape and appearance) and shared 'characters' such as feathers and scales, as well as sexual and reproductive patterns.

Since DNA is the genetic material itself, and proteins are encoded by DNA, family trees derived from DNA or protein similarities and differences

ERNST HAECKEL
1834–1919

An enthusiastic follower of Darwin, Haeckel sometimes let his enthusiasm get out of hand, bending the facts to fit the schemes of evolution he devised. Yet despite this he did a lot to promote the acceptance of evolutionary theory in Germany, among both zoologists and the general public.

have provided a test of the accuracy of older classifications. The molecular data provide quantitative information that is quite independent of morphology, and so helps to clarify which morphological traits truly reflect common descent rather than evolutionary convergence. The results have not exactly invalidated the old classifications, for many of the original taxonomists did their work well, but they have certainly helped create a clearer picture of the evolutionary process.

SERUM AND ANTI-SERUM

Several different methods have been used for obtaining molecular family trees.

The oldest, originated by G. H. F. Nuttall at Cambridge University in about 1900, is known as protein immunology. When a small amount of an animal's serum is injected into a rabbit, the rabbit's immune system makes antibodies that are highly specific for binding to the injected protein. When the original serum is mixed with this anti-serum, the mixture turns cloudy and a protein precipitate falls to the bottom of the test-tube. If serum from a different species is mixed with this anti-serum, there is less precipitate.

The amount of precipitate indicates the evolutionary relationship between the two species: the closer the relationship, the more precipitate. For example, the cross-reaction between horse anti-serum and donkey serum produces abundant precipitate, but not quite as much as horse anti-serum and horse serum. Horse anti-serum mixed with cow serum produces very little precipitate because the two groups are so distantly related. Horse anti-serum mixed with fish serum produces none at all, because of the even more distant relationship.

Nuttall found relationships among several thousand vertebrate species by this method, including the close kinship of humans and apes. However, the significance of his results was not fully appreciated at the time because the structure of proteins, antibodies and genes was unknown. Only now we know these biochemical relationships are Nuttall's discoveries appreciated.

193

DEVELOPMENT AND BODY PLAN: THE HOMEOBOX

Karl Baer, the founder of embryology, and Ernst Haeckel were struck by the way the early embryonic stages of groups like fish, amphibians, reptiles, birds and mammals resembled each other. Mammalian embryos even have gill arches like their fish ancestors! These observations led to the notion of recapitulation – that the embryo, in a few weeks or months, goes through all the evolutionary stages that its ancestors went through during millions of years. Haeckel summed up this principle in the phrase 'Ontogeny recapitulates phylogeny'.

This dictum has been assailed in modern times as a gross oversimplification of the complex process of embryological growth and development, but its kernel of truth has been vindicated by recent discovery of 'homeobox' genes that guide the common body plan of a broad range of animals.

The French zoologist Geoffroy Saint-Hilaire believed that all animals have the same basic body plan. Yet in insects the main nerve cord lies in front of the internal organs, whereas in vertebrates the spine is at the back. Undaunted, Saint-Hilaire deduced from this that vertebrates are basically upside-down insects! Unlikely as it sounds, this idea has also been supported by recent discoveries.

Insects have three body segments: the head, thorax, and abdomen. In the 1940s American biologist Edward B. Lewis began studying the genes that affect segmentation in the fruit fly *Drosophila melanogaster*, long a laboratory favourite. Drosophila reproduces rapidly, has only four chromosomes, and shows many mutations when inbred or exposed to X-rays. Lewis studied a mutation that caused a duplication of the thorax, complete with an extra pair of wings. The gene in which this mutation occurred was called bithorax. Since hundreds of genes participate in the formation of a body segment and wings, Lewis deduced that bithorax was a master gene capable of switching all the others on and off.

In the 1970s two German biologists, Christiane Nüsslein-Volhard and Eric F. Wieschaus, discovered that the ten genes controlling the fruit fly's development, the homeotic genes, are lined up on one of its chromosomes in the same head-to-tail order as the body parts they control. Furthermore, when these genes were sequenced, they all had a virtually identical DNA segment

ABOVE **A gene like bithorax can lead to mutations in a fruit fly – such as an extra pair of wings and a double thorax.**

180 base-pairs long, which translates into a protein sequence with 60 amino acids. The DNA segment is called the homeobox, the protein the homeodomain.

The homeobox is the master key to the mystery of development. The homeodomain protein binds to cellular DNA and switches on and off the process of transcription, the first step in the manufacture of the proteins that control the growth and development of the organism. About ten similar homeobox (hox) genes are found in nearly all multicellular animals. Thus the hox genes ultimately determine the shape, size, structure, and appearance of all animals, and hox genes have recently been discovered in plants and in single-celled yeasts as well. For their work on these genes Lewis, Nüsslein-Volhard and Wieschaus received the 1995 Nobel Prize for Physiology or Medicine.

Whereas the arthropods, including the insects, have about ten hox genes all in a row on one chromosome, vertebrates such as fish, birds and mammals have four rows of ten each, on four different chromosomes. Evidently the invertebrate chromosome on which the hox genes reside duplicated twice during the evolution leading to vertebrates. The rows of hox genes on different chromosomes have marked sequence similarities to each other, and to invertebrate hox genes. As in the fruit fly, the vertebrate hox genes are lined up in the order of the regions of the body they control, the gene at the front organizing the head, and so forth.

The hox genes are the molecular sculptors that shape the similar embryological forms that so intrigued Baer and Haeckel more than a hundred years ago. A very small set of related genes are almost certainly replicates of a single gene that existed in one-celled eukaryotes more than a billion years ago. These hox genes determine the physical forms of the wonderful variety of living organisms that make up our world.

Saint-Hilaire's bizarre conjecture that vertebrates are upside-down insects is also vindicated. The homeotic gene which, in an invertebrate, determines which side will be the belly, also determines which side will be the back in a vertebrate. In the dark tangles of the genome, which are only now being explored as the tropical jungles were explored in the last century, the realities being revealed are far weirder than any biologist's imagination.

ZIPPERS WITH GAPS

In the 1970s a technique called DNA hybridization was developed that allowed direct comparison between the genomes (all the genetic material) of two or more species. When DNA extracted from whole cell nuclei is heated in solution, the hydrogen bonds holding the two helical strands together are broken and the double helix 'melts' into two single strands. Single strands from two different species can then be mixed together and, as the solution cools off, some of the newly formed double helices will be hybrids having one strand of the DNA molecule from each species.

When the solution is reheated, the hybrid double helices melt apart at a lower temperature than the original double helices did, because not all the bases lined up on one strand have complementary bases on the opposite strand. The hybrids behave like zippers with gaps – gaps that correspond to the genetic differences between the two species. A difference in melting

ABOVE **Ever since Darwin started musing on the origins of the Galapagos finches, the various adaptations of birds have provided biologists with spectacular examples of evolution in action. Yet the true relationships between different groups of birds have often been masked by their adaptation for the same ways of life, and are only now being revealed by molecular studies.**

ORTHOGENESIS

Haeckel considered that the development of life through time was an inevitable unfolding, just as the embryo unfolds to become an adult. He did not accept the random nature of natural selection. Instead he thought that evolution progressed from lower to higher forms of life, with humanity at the top of the tree. This notion of inevitable evolutionary 'progress' from lower to higher forms, known as orthogenesis, is a common fallacy even today.

temperature of 1°C corresponds to a difference of about one per cent.

Charles Sibley and Jon Ahlquist have carried out DNA hybridizations on thousands of birds and discovered that a number of bird groups that were thought to be related, because of their very similar appearance, are actually quite different in their DNA. This means that the physical resemblances are due to convergent evolution rather than common descent.

For example, New World and Old World vultures had been thought to be closely related, but their DNA indicates that New World vultures are more closely related to storks. Starlings were thought to be akin to crows, but their DNA marks them as sisters to the mockingbirds. Experts were split on the owls, some thinking that their kinship lay with daytime birds of prey like falcons and hawks, while others suspected a kinship with nightjars. DNA places them with the nightjars.

LEFT **Owls are often regarded as nocturnal hawks and eagles, and you might assume they are closely related. But DNA studies show that they have quite different ancestry and are actually related to nightjars. They have evolved in similar ways to nocturnal hawks and eagles because life has given them similar problems to solve.**

RIGHT Small, insect-eating bats have many adaptations that are not shared by their bigger, fruit-eating cousins, the flying foxes. This has caused some confusion over their origins, including the suggestion that the big bat is more closely related to the monkey. Anatomical evidence is never conclusive, though, because of the changes that can be caused by evolution.

ARE BATS FLYING PRIMATES?

Bats (Chiroptera) are the only group of flying mammals, and all bats have many anatomical features in common, including wings adapted from ancestral hands. It is logical to deduce that such a complex adaptation evolved only once, and that all bat species are more closely related to each other than to any non-flying group of mammals. This traditional view was challenged, however, by neuroanatomist J. D. Pettigrew, who observed similarities in the nerves of the eye and midbrain between the big bats or flying foxes (Mega-chiroptera) and primates like monkeys. The small bats (Microchiroptera) apparently do not have these nerve connections.

Pettigrew therefore proposed that mega-chiropterans are more closely related to primates than to microchiropterans. If he was actually correct, all the features of wing structure and powered flight shared by the big bats and little bats had evolved independently from two different ancestors.

To resolve this controversy, nucleotide sequences were compared from three different molecules: ribosomal RNA, mitochondrial DNA, and the nuclear eta-globin gene, one of the many haemoglobin genes. All three sequences supported the traditional allying of the big bats with the little bats, and none confirmed Pettigrew's postulated relationship between the megachiroptera and primates. Immunological comparisons of serum albumin of big bats and little bats also strongly favour the traditional view.

This is just one of many examples in which molecular family trees derived from different genes or proteins, and carried out by different researchers, have agreed with each other, while morphology-based family trees emphasizing selected anatomical features have arrived at widely divergent conclusions.

A THIRD FORM OF LIFE

Ribosomes are the organelles in the cells that translate messenger RNA (mRNA) into proteins. In fact ribosomes are made of RNA and proteins. This ribosomal RNA (rRNA) is remarkably stable, showing relatively few mutational differences between species tens or hundreds of millions of years distant from their common ancestor. The exact structure of rRNA is vital to the transcription process in the ribosomes, and so most mutations would interfere with transcription.

The relative stability of rRNA allows it to be used to examine the genetic relationships of organisms that, on the face of it, seem to be barely related at all. Starting in the 1960s, C. R. Woese compared rRNA sequences of bacteria and higher organisms. At that time, biologists divided the living world into two parts: the prokaryotes, whose cells were without nuclei, and the eukaryotes whose cells did have nuclei. The prokaryotes were the bacteria alone. The eukaryotes are everything else, including fungi, plants, animals, and protists (single-celled organisms with large nucleated cells).

Comparing the rRNAs of different kinds of bacteria, Woese and his co-workers made a totally unexpected discovery. Some odd bacteria that live in extreme environments – either without oxygen, at high salt concentration, at high temperatures, or that generate methane instead of carbon dioxide – were as different from normal bacteria as the bacteria are from eukaryotes. So instead of being split into two major parts the living world was divided into three parts: the eukaryotes, the bacteria, and the odd bacteria-like organisms that Woese named archaebacteria, and are now called archaea. Because bacteria and archaea actually look alike under the microscope, one of the major kingdoms of life on Earth had been completely overlooked for centuries!

THE RNA WORLD

Although all living cells depend on processes involving **DNA**, **RNA** and proteins, numerous viruses have **RNA** rather than **DNA** as their active ingredient. It was discovered in the mid-1980s that **RNA**, unlike **DNA**, can also perform some of the enzymatic functions needed for replication, which in principle might make proteins unnecessary. So origin-of-life buffs now speculate that a simpler 'RNA world' preceded the DNA-plus-protein world we live in.

RNA might have been able to replicate and evolve without specialized proteins. As a matter of fact, **RNA** viruses like the human immunodeficiency (**HIV**) virus that causes **AIDS** evolve at about a million times the rate of nuclear **DNA**. These retroviruses reverse the normal cellular process of transcribing **DNA** into **RNA**: they multiply by transcribing **RNA** into **DNA**, which then hijacks the cellular machinery into making more viral **RNA**.

It is therefore easy to imagine a scenario in which a bolt of lightning flashing through the steamy ammoniac haze of the Hadean atmosphere produced a few chains of **RNA**, which then proceeded to replicate themselves on the warm intertidal clay. Eventually the evolving virus-like **RNA** organisms got around to transcribing **DNA**, which was a much more efficient replicator and outcompeted its parent **RNA**.

At present this story of the **RNA** world is purely speculative, but perhaps we'll get lucky and find biochemical fossil evidence of the **RNA** world – a community which, like Atlantis, may have once lived but now is lost beneath the sea. We don't have any identifiable descendants of Atlantis, but we do have millions of species of **DNA**-based and **RNA**-based organisms on this third planet from a medium-sized star.

Some kind of life or proto-life was gestated in those pregnant planetary exhalations four billion years ago, and in our present state of knowledge (and ignorance) **RNA** is the most likely contender.

MESSENGER RNA

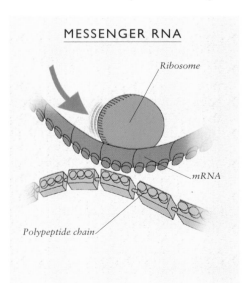

Ribosome

mRNA

Polypeptide chain

LEFT **RNA** (ribonucleic acid) is just as important as **DNA**. Messenger **RNA** contains the instructions for creating the proteins which are themselves needed by the body to survive. It is used by cellular organelles called ribosomes to build the long chains of amino acids that form these proteins.

ABOVE **The simplest forms of life are bacteria and archaea: single cells with no nuclei, known as prokaryotes.**

Molecular evidence is accumulating that eukaryotes are more closely related to the archaea than they are to the bacteria. To put it another way, we humans are more closely related to the archaea than the archaea are to the bacteria. This is despite the fact that both types of micro-organisms are single-celled and lack nuclei, and we are multicellular eukaryotes with big brains – the self-appointed Masters of the Universe. This is just one of many bizarre new slants on life to come out of the molecular exploration of the genome.

Bacteria provide the ultimate illustration of the inadequacy of morphology alone for building family trees. Bacteria are all roughly the same size and can take on only a few different shapes, mostly spheres (cocci) or cylinders (bacilli). These may cluster together in various patterns: streptococci tend to form linear chains, staphylococci form grape-like bunches. So nothing in their appearance told the taxonomists how the tens of thousands of living species were related to each other, and until Woese did his work no-one had a clue that the archaea constituted a third form of life.

WHALE TALES

The origins and relationships of whales, dolphins, and porpoises – the cetaceans – have always mystified biologists. These warm-blooded, intelligent and social mammals are so well adapted to their aquatic environment that, anatomically, there are few clues to their evolutionary affinities with ancestral land animals. But several comparative anatomists and paleontologists have suggested a connection with the ungulates, the hoofed mammals, and this suggestion has been backed up by molecular studies.

The molecular evidence for a rather close relationship between cetaceans and artiodactyls (even-toed ungulates) has been growing since the 1960s. The connection was first suggested by immunology, confirmed in the 1980s by protein sequence data, and strongly reinforced in the 1990s by mitochondrial DNA sequences. The artiodactyls include several different subgroups: pigs, camels, hippopotami and ruminants (such as giraffes, cows, gazelles and deer). In fact DNA and immunological studies indicate that the cetaceans are slightly more closely related to the hippopotami than to the others. The hippopotami would certainly be an appropriate cetacean ancestor, since they are already semi-aquatic with shortened legs and nostrils that have migrated upward.

Traditionally cetaceans have been divided into two groups: the echo-locating, toothed odontocetes and the filter-feeding mysticetes. The odontocetes include the dolphins, porpoises and sperm whales; the mysticetes include

the baleen whales (the humpback, bowhead, right, grey, blue, fin, sei, Bryde's and minke whales). But analysis of RNA and DNA sequences has challenged this traditional classification, connecting the sperm whale – even though it is a toothed echo-locator – more closely to the baleen whales than to the dolphins and porpoises.

This revised family tree, would imply that the common ancestor of odontocetes and mysticetes was a toothed echo-locator, and that both teeth and the ability to echo-locate were lost in the baleen lineage but retained in the sperm whale.

MOLECULAR CLOCKS

In the 1960s techniques became available for obtaining the amino acid sequences of proteins. These sequences enabled investigators to count the amino acid differences between a given protein in two or more species. Consequently, Emile Zuckerkandl and Linus Pauling sequenced the haemoglobin – the red blood pigment – of numerous vertebrate species. They noticed that the number of amino acids that differed between any two species seemed to be proportional to the time since these species had diverged from a common ancestor, as estimated from the fossil record. They were the first to suspect that mutation rates may serve as 'molecular clocks', providing a way of estimating species' times of origin that is quite independent of the fossil record.

The notion that biological molecules like proteins and DNA might undergo mutations at fairly regular rates over long periods of time

BACTERIA, ARCHAEA AND EUKARYOTES

From the fossil record it appears that bacteria or archaea had the Earth to themselves for about three billion years after life began. Both are prokaryotes, which lack a cell nucleus and membrane-bounded organelles, and their genetic information takes the form of a single circular **DNA** molecule lying in the cytoplasm. In bacteria the genes are densely packed, with no intervening sequences (introns) of non-coding **DNA**. Archaea do have some introns, and in that respect they are more like eukaryotes than are the bacteria.

The earliest known prokaryote micro-fossils, dating from 3.8 billion years ago, resemble modern photosynthetic cyanobacteria in their form and behaviour. This means that their basic form has persisted unchanged for billions of years. Meanwhile the evidence of molecular clocks indicates that bacteria and archaea probably diverged about 2.5 billion years ago.

Eukaryotes are much larger than prokaryotes and have a membrane-bound nucleus. About half the genes of the intensively studied bacterium Escherichia coli match genes that are found in the eukaryotes. So eukaryotes must be derived from prokaryotic ancestors.

Nearly all eukaryotes are symbiotic composites, each consisting of a big eukaryotic cell and a number of bacteria-derived mitochondria and – in plants – chloroplasts. Molecular sequence comparisons show that mitochondria are descendants of a particular line of bacteria called alpha proteobacteria, and chloroplasts are descendants of photosynthesizing cyanobacteria. These organelles were acquired at some early stage of eukaryotic evolution, and a few primitive eukaryotes, like the parasitic protist Giardia which diverged very early from the eukaryotic line, do not have them. Their importance to the nature of life on Earth is pivotal, for mitochondria provide the cell with energy through utilization of oxygen, and chloroplasts contain the chlorophyll that converts the Sun's energy to sugar molecules.

ABOVE **DNA** studies have shown that there is a relationship between cetaceans and artiodactyles which means there is a common link between whales, dophins and porpoises, and cows, pigs and hippopotami.

– during tens or hundreds of millions of years – was totally unexpected before the mid-1960s. Investigations of four types of protein molecule – haemoglobins, histones, cytochrome c, and fibrinopeptides – indicated that these molecules had changed at regular rates over long periods, but the rate of change in each protein is very different.

R. F. Doolittle derived family trees by comparing the amino acid sequences of fibrinopeptides. Like haemoglobins, fibrinopeptides have a fairly constant clock-like rate of change, but the family trees of ungulates (cattle, deer and their relatives) and other vertebrates showed that fibrinopeptides accumulate mutations more rapidly than haemoglobins over the same evolutionary time periods. Comparisons with other proteins showed dramatic differences. Fibrinopeptides undergo amino acid substitutions at 900 times the rate of histones, for example. So these biomolecules behave rather like the radioisotopes used for rock dating, which decay with different half-lives depending on their nuclear structure. Students of evolution quickly realized that if molecular clocks were even approximately accurate, they would be as great a boom to evolutionary understanding as radioisotopes had been to dating geological strata.

HOW DOES THE
MOLECULAR CLOCK WORK?

There is still controversy over why biomolecules behave as clocks, how accurate they are and what factors – such as generation time and sex – affect the rates at which they run.

It is generally accepted that the reason why molecular clocks work at all is that the great majority of mutations are neutral, and apparently have no significant effect on the descendant individual. In the 95 to 99 per cent of the eukaryotic genome which consists of non-coding DNA, the mutation rate is five to ten times faster than it is in the one per cent which codes for proteins. Virtually all the non-coding mutations are neutral – in the sense that they do not change the physical characteristics of the organism – whereas some but not all mutations in the genes have deleterious effects and are therefore eliminated over time.

Even in the genes that code for proteins, many mutations are 'silent'. A change in the third position of a three-letter codon on the mRNA strand often does not change the amino acid for which it codes. For instance, CUU, CUC, CUA and CUG are all codons for the amino acid leucine. Therefore any third-position mutation would still produce leucine, whereas mutations in the first and second positions (C and U) would instruct the ribosome to build a different amino acid into the protein. These 'synonymous' substitutions in the third codon position are observed much more often than the 'non-synonymous' substitutions in the first and second positions.

An even more definitive demonstration of the rate difference between coding and non-coding sequences is provided by 'fossil' sequences within the genome known as pseudogenes. Some once-active genes may become inactive 'dead genes' due to mutations in the flanking sequences that control transcription. These pseudogenes continue to replicate from generation to generation but no longer encode proteins, and therefore have no significant effect on the organism. For example, a number of haemoglobin pseudogenes have been identified. It is therefore possible to compare the mutation rates in haemoglobin genes and haemoglobin pseudogenes by matching sequences in different species, such as humans and horses. In many such comparisons, the pseudogenes have undergone about five times as many mutations as the functional genes.

Most mutations are thought to be due to 'copy errors' that occur during the process of the replication of the reproductive cells. So it has been conjectured that molecular evolution should proceed faster in species like rodents that have short generation times and reproduce more frequently than long-lived species like humans and elephants. Yet despite considerable work aimed at either proving or disproving this generation-time effect the matter is still not resolved. Likewise, since the male reproductive cells undergo many more replications than the female ova, it has been hypothesized that males contribute more mutations to the gene pool than females do. The evidence supporting and contesting this hypothesis is also not yet definitive.

Do the same genes evolve at the same rate in different groups of organism? This has been one of the most persistent issues in the ongoing molecular clock debate for the past three decades, and it is far from being completely settled. The two extreme views have been represented by Morris Goodman and Allan Wilson. Goodman and his colleagues at Wayne State University have long denied any kind of regular mutation rate over long time

ABOVE **The discovery of the molecular clock in the 1960s has allowed biologists to estimate when species' originated, thus providing us with further information on the evolutionary process.**

periods. Conversely the late Allan Wilson of the University of California at Berkeley maintained the existence of a universal clock of synonymous substitutions, which is about the same for all living organisms.

As is so often the case, the truth appears to lie somewhere between these polar views. Overall there seem to be fairly consistent rates of mutation for homologous genes and for non-coding DNA, but with many exceptions observed in specific lineages. Rodent DNA for example, seems to evolve faster than primate DNA. Within the primates hominoid DNA probably evolves somewhat more slowly than monkey DNA, and human DNA is perhaps the slowest of all. This remains an active field of research.

Whatever the actual rates of mutation, the number of mutational differences between two species always increases with the passage of time. This molecular divergence is often the only indicator of the time elapsed since common ancestry, particularly with organisms like bacteria for which there is little or no fossil record. Within a given lineage such as carnivores, or for a given molecule such as haemoglobin, the rates of change seem to be sufficiently consistent to justify using them to draw family trees with time estimates as coordinates. It helps if several different molecular clocks are used and averaged. As a result the molecular clock concept has greatly enriched and clarified our understanding of the timescales over which the Tree of Life has ramified its many branches.

MULTI-CELLULAR LIFE

Molecular clocks time the origin of multi-cellular animals to about 720 million years ago, although as fossils they make their earliest appearance in rocks of the Vendian Period of the late Precambrian, about 600 million years ago.

This multi-cellular debut occurred 50 million years before the great Cambrian explosion of metazoan life forms. In a period of only 10 million years – a geological instant – 20 or 30 of the modern phyla and more than 50 new orders emerged: sponges, jellyfish, flatworms, round-worms, trilobites and a variety of segmented soft-bodied creatures, many of them now extinct.

Molecular comparisons attest that all the different multicellular forms still found on Earth arose from a single common ancestor. One defining molecular characteristic of the metazoans is the protein collagen. This has a triple-helical structure, forms long rope-like fibres and is a principle component of sponges, shells, skin, bones and teeth. Collagen is an important factor in the formation of hard body parts such as protective armour, biting jaws and skeletons.

By about half a billion years ago the body plans of virtually all the types of animals that live today were established. Their increased complexity is reflected in the increase in genome size, from four million base pairs in bacteria (4000 genes) to 15 million in single-celled eukaryote yeasts (6000 genes), 100 million in nematode worms (13,000 genes) and 3000 million in mice and men (80,000 genes).

Yet within the dazzling array of novel morphologies that emerged in the Cambrian era, the genetic molecular mechanisms remained surprisingly conservative. Though organisms grew larger, developed appendages and nervous systems and found new ways of feeding, most of the bacterial signalling systems, transcriptional processes and genes for structural proteins continued to function as they had in prokaryotes – the organisms that first evolved on Earth.

Chimpanzee
The chimpanzee's genetic material is less than 1% different from our own

Gorilla
Gorillas' genetic material is 99% the same as ours

Human
The similarity in genetic material shows that we humans must have evolved from the same ancestor as chimpanzees and gorillas

LEFT **Despite our very different appearance we are more closely related to chimpanzees and gorillas than we might like to think. Our genetic material is almost identical, and chimpanzees have a lot more in common with us than with the other apes – the orang-utan, baboons and gibbons.**

APES AND HUMANS

One of the first applications of the molecular clock concept – an application that split the community of anthropologists into warring factions – was the research of Vincent Sarich and Allan Wilson that indicated humans and apes diverged from a common ancestor recently.

Using an immunological technique that was both more sensitive and more quantitative than that used by Nuttall at the beginning of the century, Sarich and Wilson compared the serum albumin of humans with the albumins of the African apes (the chimpanzee and gorilla), the Asian apes (the orang-utan and gibbons), and other primates (particularly Old World and New World monkeys).

They found that human, chimpanzee, and gorilla albumins were nearly identical, with only one per cent difference between any pair. Orang-utan and gibbon albumins were twice as different from these three (about two per cent) as those of the human-chimpanzee-gorilla threesome were from each other. The albumins of these five hominoids were about six per cent different from those of Old World monkeys such as baboons.

The fossil record indicated that apes and Old World monkeys had diverged from a common ancestor about 30 million years ago. So assuming that albumin, like other proteins, underwent a steady rate of change, it must have taken 30 million years to undergo a six per cent change. Therefore the one per cent difference between human, chimpanzee and gorilla albumin would have developed over one-sixth of 30 million years – namely, five million years.

So Sarich and Wilson concluded that humans, chimpanzees and gorillas had a common African ape ancestor living five million years ago. This conclusion unleashed a storm of protest and debate that went on for more than two decades. It pitted the new practitioners of molecular evolution against some of the veteran paleontologists, who were convinced by the fossil record known in the mid-1960s that early hominids had walked in Asia and Africa as long as 20 million years ago.

A TEST CASE FOR MOLECULAR CLOCKS

Exhibit A in the case against molecular clocks was a fossil known as *Ramapithecus*, found in rocks laid down some 20 million years ago in the Miocene Epoch. Jaws and teeth of *Ramapithecus* excavated from sites in India (now Pakistan) were interpreted by their discoverers as resembling human jaws and teeth. No pelvis bones or limb bones of this Miocene hominoid had been discovered at that time, but its sponsors were so confident the dental apparatus was hominid (on the human line), that *Ramapithecus* was routinely depicted in articles and textbooks as an upright walking member of the human lineage. Similar fossils were also found in African Miocene deposits, and this discovery seemed to leave wide open the issue whether the human line originated in Asia or Africa.

The molecular evidence of Sarich and Wilson's albumin studies implied that *Ramapithecus* lived before the common ancestor of humans and modern African apes, and therefore could not be considered hominid. The paleontologists could not accept this, and the disputes over the real place of *Ramapithecus* in hominid evolution – and its relationship to the common ancestor – became quite heated. Most paleontologists were adamant that *Ramapithecus* was directly related to humans.

Most such disputes in science are settled by the accumulation of more evidence and new methods, and the *Ramapithecus* controversy was no exception. From the molecular side, data rapidly piled up showing that additional protein molecules besides albumin, as well as DNA, confirmed the very close relationship between humans and African apes. This implied a divergence closer to five than to 20 million years.

Even the paleontological data supported this view. Fossils of early hominids, the australopithecines of eastern and southern Africa that lived three to four million years ago, anatomically resemble chimpanzees in brain and body size, suggesting an ape ancestor in the not-too-distant past. Finally, additional *Ramapithecus* fossil cranial and limb bones were discovered in

Baboon
Baboons and other monkeys may look like apes but genetically they are quite different

Gibbon
Although apes, gibbons are not as directly related to us as gorillas and chimpanzees

RIGHT **The baboons and other old world monkeys are descended from the distant common ancestors of apes and humans – but they are now not directly related. To an ape, the idea that it might just be related to a monkey would probably be as disturbing as the idea that we are related to apes.**

Pakistan. The cranial bones included teeth that were different from either modern humans or modern apes, and the limbs were those of a tree-living ape rather than a bipedal human.

So the predictions of the molecular clock hypothesis held up. *Ramapithecus* proved not to be a hominid, and hominid fossils dated to some five million years ago resemble African apes more than any other primate.

MITOCHONDRIAL DNA: A FAST CLOCK

Within the cells of animals, plants, and fungi are many small organelles called mitochondria, which serve as power sources for the cell by promoting the oxidation of carbohydrate fuel. The ancestors of the mitochondria were bacteria that entered into a symbiotic relationship with eukaryotic cells more than a billion years ago. The mitochondrial genome is separate from the nuclear genome, and has been found to undergo mutations at five to ten times the average rate of the nuclear genome. It can therefore be used as a 'fast clock' to time recent evolutionary events.

A mammal's mitochondrial genome consists of a circular double-stranded DNA molecule like that of bacterial DNA, about 16,000 base pairs long. Mitochondrial DNA (mtDNA) is transmitted exclusively through the maternal line in most species because mitochondria reside in the cytoplasm. A sperm has almost no cytoplasm, being basically a cell nucleus propelled by a whip-like flagellum, so all the mitochondria of the fertilized zygote are derived from the maternal ovum. And although mtDNA consists of about 37 genes, these gene sequences do not recombine in the way nuclear genes do because there are no paternal mtDNA genes with which to recombine. So mtDNA remains unchanged in the maternal lineage from generation to generation, except for mutations. An individual might very well have mtDNA identical to that of his or her great great grandmother, whereas all of the 70,000 or so nuclear genes would be scrambled up with the contribution of both the female and male progenitors of the intervening generations.

Because mtDNA evolves relatively rapidly but does not recombine, it has proved useful in defining geographical sub-populations among humans, as well as many other species of animals and plants. In particular, studies of mtDNA in different human races have provided the most conclusive evidence to date about the emergence of the modern human species *Homo sapiens*.

THE MISSING LINK

Ape   Missing link   Human

WHICH APE IS OUR CLOSEST RELATIVE?

The earliest molecular evidence from protein immunology and protein sequences was not able to determine the closest pair among humans, chimpanzees, and gorillas. In 1984 this problem was tackled by ornithologists Charles Sibley and Jon Ahlquist, who had used DNA hybridization to clarify bird relationships. Sibley and Ahlquist proceeded to hybridize DNA from humans, two species of chimpanzees, gorillas, orang-utans, gibbons and baboons. Their results were expressed as differences in melting temperatures of the DNA hybrids between humans and other primates. The human-chimpanzee DNA melting temperature came out lowest at 1.8, while the hybrid with gorilla DNA was 2.4, orang-utan 3.6, gibbon 5.2 and baboon 7.7. These numbers are roughly equal to the percentage sequence differences between the species.

These experiments clearly demonstrated that our closest living relatives are chimpanzees, a kinship that is consistent with the chimpanzee-like anatomy of the earliest known hominids. It is also consistent with observations of ape behaviour, especially the early pioneering studies of Jane Goodall on chimpanzees in Gombe, Tanzania, and later observations by Japanese primatologists Toshisada Nishida and Tokayoshi Kano on the rarer pygmy chimpanzee. The ability of these apes to make and use tools, to communicate with gestures much like our own, and to maintain close social relationships has revealed that many

LEFT **For years evolutionists looked for the 'missing link' between apes and humans. They thought they had found it in** *Ramapithecus,* **but molecular studies have knocked this idea out of court.**

ABOVE **We are not descended from chimpanzees, although we are closely related. We are variations on the same theme, with common ancestors that walked the Earth maybe five million years ago.**

of Sibley and Ahlquist: human and chimpanzee were the closest pair and gorillas were more distant from both. Subsequently more than a dozen DNA sequence comparisons by independent researchers have shown the same pattern of relationships. Statistically, these data give a probability of greater than 99 per cent that the human-chimpanzee connection is correct.

characteristics we once thought were uniquely human are shared with these hominoid relatives, which also share more than 98 per cent of our genes.

In some respects the human-chimpanzee pair contradicts common sense, for a chimpanzee and gorilla look more like each other than either looks like a modern human. But numerous additional studies of DNA have confirmed the chimp-human connection. This strongly suggests that the common ancestor of chimpanzees and humans was a knuckle-walking ape very much like a chimpanzee. The oldest possibly hominid fossil from Aramis, Middle Awash, Ethiopia is dated to about 4.4 million years ago. These fossils have not been fully described at the time of writing, but preliminary reports show a remarkable similarity between the teeth of this oldest hominid and the teeth of the pygmy chimpanzee.

Inevitably the findings and the methods of Sibley and Ahlquist met with criticism from anthropologists who assumed, for one reason or another, that the closest relationship was chimpanzee-gorilla or human-gorilla. There was considerable resentment in some quarters that this long-standing puzzle in human evolution had been resolved, or was professed to be resolved, by ornithologists who knew little or nothing about primates. So strong were feelings on the subject that Sibley and Ahlquist were accused by several prominent anthropologists of falsifying their data. Human evolution arouses violent emotions, even among those dedicated to studying it!

When an independent DNA hybridization of hominoids was carried out (this time by entomologists) the results were identical to those

ABOVE **Fossil remains of early hominids are scarce and confusing, for they suggest that several different species may have existed alongside each other in the distant past. Their fragmentary remains tell us little about their relationship to each other, and do not preserve the DNA that could settle the arguments.**

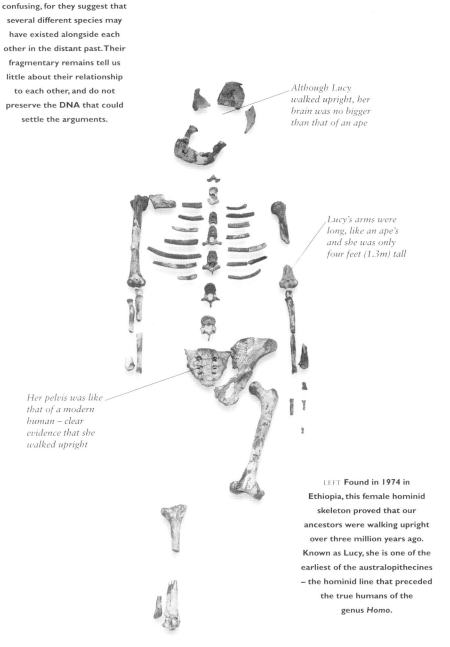

Although Lucy walked upright, her brain was no bigger than that of an ape

Lucy's arms were long, like an ape's and she was only four feet (1.3m) tall

Her pelvis was like that of a modern human – clear evidence that she walked upright

LEFT **Found in 1974 in Ethiopia, this female hominid skeleton proved that our ancestors were walking upright over three million years ago. Known as Lucy, she is one of the earliest of the australopithecines – the hominid line that preceded the true humans of the genus *Homo*.**

VERTEBRATE EVOLUTION

The first vertebrates, the jawless fish, appeared in the Ordovician Period about 450 million years ago. Hagfish and lampreys are the living relics of these ancient forms, and lampreys almost identical to those living today have been found in fossil shales of the Devonian Period 350 million years old. Fish with proper articulated jaws also proliferated during the Devonian, and for the first time true bone appeared, providing more rigid support than the cartilaginous skeletons of the jawless fishes and sharks.

Some of these fish had fleshy, muscular fins which were the precursors of legs. Others had an internal air sac known as a swim bladder – still a feature of most bony fish – and in the lungfish this evolved into a primitive lung with which to breathe air. These adaptations were – almost literally – the first steps towards leaving the sea and living on land. Exactly which kind of fish made this transition is the subject of much debate, with some experts favouring the ancestral lungfish and others the progenitors of the coelacanth, a primitive fish with fleshy fins and a mostly cartilaginous skeleton. As with so many such controversies molecular studies have helped clarify the issue, for comparisons of many genes indicate that the lungfish is the more likely candidate.

ABOVE **All land vertebrates, from frogs and lizards to eagles and elephants, can trace their ancestry back to primitive forms of fish. We still carry the remains of gills in our bodies, as the tiny bones of the inner ear.**

LEFT **The Mesozoic Period is best known for the dinosaurs that dominated the planet during this time. However, there were also many plants, birds, mammals and insects in existence that we would still recognize today.**

ABOVE **The vertebrate body plan has proved to be one of the most adaptable products of evolution, capable of extraordinary transformations into such unlikely creatures as flying fish and birds that 'fly' underwater.**

The first land vertebrates were amphibians, which with lungs, legs, thick skins and strong skeletons were able to survive for long periods on land – even though they still had to return to the water to lay their eggs. They flourished during the Permian Period about 250 million years ago. The beginning of the Triassic Period, 25 million years later, marked the start of the Mesozoic Era – known as the Age of Reptiles because of the dominance of the dinosaurs. During this era birds, mammals, flowering plants, and many modern insects appeared for the first time. These species survived the great extinction at the close of the Cretaceous Period 65 million years ago. When the dinosaurs were eliminated, the mammals gradually came into their own and now are the dominant vertebrates on land all over the world.

Modern elephants are now restricted to the tropics, although they were once widespread throughout the world

The powerful mammoth was adapted for life in the ice age, with a thick coat of hair to keep out the cold

DINOSAUR DNA

One of the lingering questions today is whether birds are the descendants of dinosaurs. Most paleontologists think so, and of course some dispute it. Unfortunately live dinosaurs are not available, so we cannot directly compare the DNA and proteins of birds and dinosaurs. If we could extract DNA and protein molecules from fossils, we could greatly extend our knowledge of both living and extinct species.

The novel and movie *Jurassic Park* depicts dinosaurs recreated from fossil DNA. The author Michael Crichton develops an ingenious scenario in which mosquitoes that have been fed on dinosaur blood are preserved in amber, making it possible for modern scientists to extract and clone dinosaur DNA. Crichton based his novel on scientific reports of the extraction of identifiable DNA from insects trapped in amber millions of years old. But can DNA survive so long?

Fossil molecules have certainly been identified. The first, in 1980, was albumin from the muscle of a frozen baby Siberian mammoth called Dima. The albumin was analysed by an ultrasensitive immunological method called radioimmunoassay, and found to differ from the albumins of African and Indian elephants by one per cent. The albumin molecular clock ticks at one per cent every five million years, so this implied that the three elephant species – mammoth, African, and Indian – had a common ancestor about five million years ago. Many paleontologists had considered the mammoth to be more closely related to the Indian elephant because of similarities in their teeth, but radioimmunoassay showed that the three species were equidistant. Subsequently, radio-immunoassay analysis was successfully applied to the remains of two more extinct species: the American mastodon and Steller's sea cow.

SEA COWS AND ELEPHANTS

Sea cows are the elephant's closest living relatives and radioimmunoassy analysis made it possible for the first time to construct a molecular family tree including two extinct elephant species, two living elephant species, an extinct sea cow and three living sea cows.

The first successful extraction and cloning of fossil DNA was in 1984. The material came from the skin of a quagga, an extinct member of the horse family, or equines. In the early nineteenth century, quaggas roamed the South African plains in vast herds, like bison in North America. Hunters slaughtered them, farmers destroyed their habitat, and the last quagga died in the Amsterdam zoo in 1887.

The quagga was striped on its front half like a zebra and chestnut-toned like a horse on its hindquarters. Experts on horse evolution came up with three different theories about the quagga's relationship to other equines, based largely on its stripe pattern, bones and teeth. Theory One stated that the quagga's closest relative was the domestic horse. Theory Two proposed that its closest relative was the plains zebra. Theory Three maintained that it was equally closely related to all three living African zebra species: the plains, mountain and Grevy's zebras. This is a classic example of the inability of morphology alone to establish genetic relationships.

In the 1980s the quagga conundrum attracted the interest of two research teams in different laboratories, one concentrating on protein analysis and the other on the newly refined techniques for sequencing and amplifying DNA. Using these techniques a team from the University of California, Berkeley, succeeded in extracting and cloning a short stretch of mitochondrial DNA from a quagga skin preserved in a museum. At the same time another team, from the University of

LEFT **Zebra or horse? The now-extinct quagga might have been either, but genetic material preserved in quagga skin has enabled molecular biologists to work out its family tree, and even suggest ways of bringing it back to life.**

The mammoth and the modern elephants all had a common ancestor, and if man had not wiped it out the mammoth would still be with us today

ABOVE **Modern elephants once had a spectacular relative: the mighty mammoth. By sheer luck several mammoths have been found preserved in the permafrost of the Siberian tundra. DNA from these remains has been used to work out exactly how mammoths and elephants were related.**

California, San Francisco, did radioimmunoassay on serum proteins extracted from another quagga skin. The DNA team from Berkeley compared quagga DNA with that of other members of the horse family, and the radioimmunoassay team from San Francisco made similar comparisons with serum proteins.

Both techniques produced very similar family trees of the quagga and other equines. Both support Theory Two – that the quagga's closest relative is the plains zebra. In fact, the quagga is so close to the plains zebra that the two were probably subspecies rather than distinct species. If so, then most or all of the extinct quagga's genes live on, in the plains zebra.

Proceeding on this assumption, the South African Museum in Cape Town instituted a selective breeding programme with the goal of bringing the quagga back from extinction. Plains zebras that resemble quaggas, in having little or no striping on their hindquarters, are being inbred at a nature conservation station. Presumably the progeny will be genetically almost identical with the quagga. If this experiment works, it will be the first time in evolutionary history that an extinct species has been successfully restored to life.

WILL THE DINOSAUR LIVE AGAIN?

The quagga story makes several significant points. First, biomolecules – even very ancient ones – may survive in fossil material. Second, with modern techniques these molecules may be sufficiently intact to yield information about species relationships. Third, in extinct as well as living organisms, morphological comparisons do not always give us a correct understanding of genetic relationships. Finally, molecular methods are a powerful tool in conservation genetics, in studying groups of animals on the edge of extinction and occasionally, as with the quagga, helping them back from the brink.

Returning to *Jurassic Park*, it would be thrilling to recover dinosaur DNA or proteins and compare these with molecules from birds and other reptiles. But as yet no-one has succeeded in recovering identifiable macromolecules from dinosaur fossils, which are of course at least 65 million years older than the sub-fossil remains of quaggas or even mammoths, and lack all soft tissue. Some researchers have reported recovering DNA from insects in amber as old as 130 million years, which would put them in the age of the dinosaurs, but others have been unable to replicate these results. So it is not at all certain that macromolecules are able to survive for such a long time, even in amber.

The study of ancient molecules, however, is still in its adolescence, if not its infancy. Techniques for detecting and sequencing DNA and protein molecules are improving all the time and we may one day recover genetic material from dinosaurs.

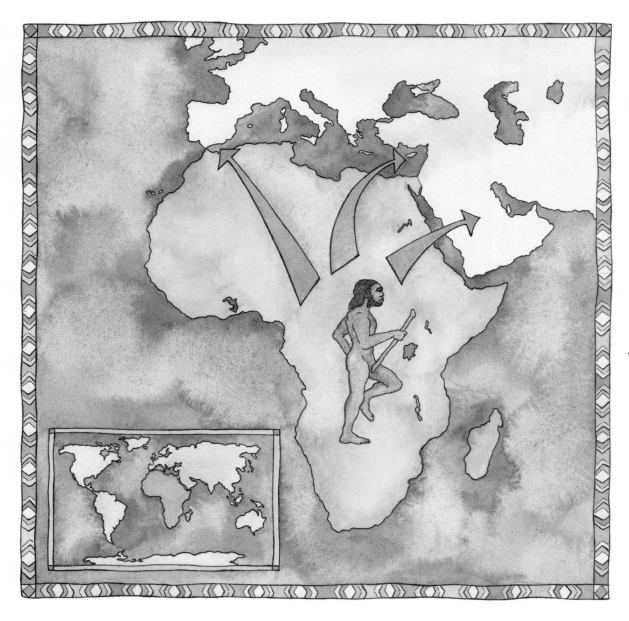

MARY LEAKEY
b. 1913

Mary Leakey, one of the great paleoanthropologists of the twentieth century, famous for the many discoveries of hominids which she made with her husband. In 1976, she discovered 3.75 million-year-old hominid footprints in a layer of volcanic ash at Laetoli, Tanzania.

LEFT The discovery of fossils has provided us with evidence to show that the genus *Homo* was in Africa some 2.5 million years ago and that different species then migrated to other parts of the world.

THE ORIGINS OF MODERN HUMANS

By the 1960s hominid fossils were accumulating in South Africa through the efforts of Raymond Dart, Robert Broom and John Robinson. In East Africa, Mary and Louis Leakey had just begun their many spectacular discoveries at Olduvai Gorge, Tanzania. In 1959 the Leakeys found at Olduvai the remains – including a skull – of *Zinjanthropus boisei*. Potassium-argon dating established this human ancestor – the first hominid to be reliably dated – at 1.75 million years old. The following year, the Leakeys made a further find – the remains of another early hominid, which came to be known as *Homo habilis*, or 'handy man'. *Homo habilis* is the earliest tool-making hominid and had a relatively large brain. There is evidence that they used stone tools to cut hides for clothes and meat for eating and built crude shelters. The Leakeys' discoveries established East Africa as a possible birthplace for humans.

Since then, new fossil discoveries from several locations in Ethiopia, Kenya, Tanzania, Chad, Malawi and South Africa have uncovered several species of early hominids. All classified in the genus *Australopithecus*, they date from between two and four million years ago. Anatomically they have small brains like chimpanzees, but their leg bones, pelvises and footprints indicate that they walked upright on two legs. No evidence for hominids living more than five million years ago has been found in Africa, and no fossils dating back further than two million years have been discovered anywhere else in the world.

Between two and two-and-a-half million years ago the genus *Homo* seems to have made its appearance in Africa, and there was probably more than one species. Fossils from Java have also been dated to nearly two million years ago; these appear to be *Homo*, though the species is not well established. These remains testify that hominids migrated out of Africa by two million years ago and may have dispersed in several waves by way of Arabia, Gibraltar or the Levant.

The fossil and archaeological records in Europe, Asia, and Africa continue to expand, with new discoveries dated from two million years ago to as recently as 150,000 years ago. By this time hominids covered much of the Old World, in the form of at least two species, *Homo erectus* and *Homo heidelbergensis*.

BELOW **Darwin's view of his own species was revolutionary, in that he did not accept the prevailing opinion that humans were the pinnacle of evolution.**

ABOVE **Some three and half million years ago footprints were left in damp volcanic ash by two of our most distant ancestors. Discovered in 1978 at Laetoli in Tanzania, they show the early *australopithecines* had feet that were already adapted for walking upright.**

DARWIN AND HUMAN EVOLUTION

The work of Charles Darwin profoundly affected the study of the human species in two ways. First, he implied in the *Origin of Species* that human origins could be explained by the workings of the natural world rather than by divine intervention. Second, in his later work *The Descent of Man* (1871), he stated that human origins could be traced to Africa because of our anatomical similarities to the African apes – the chimpanzee and gorilla. Darwin's deductions relied on comparative embryology and anatomy, for at the time there were no human fossils from Africa to confirm his conclusions.

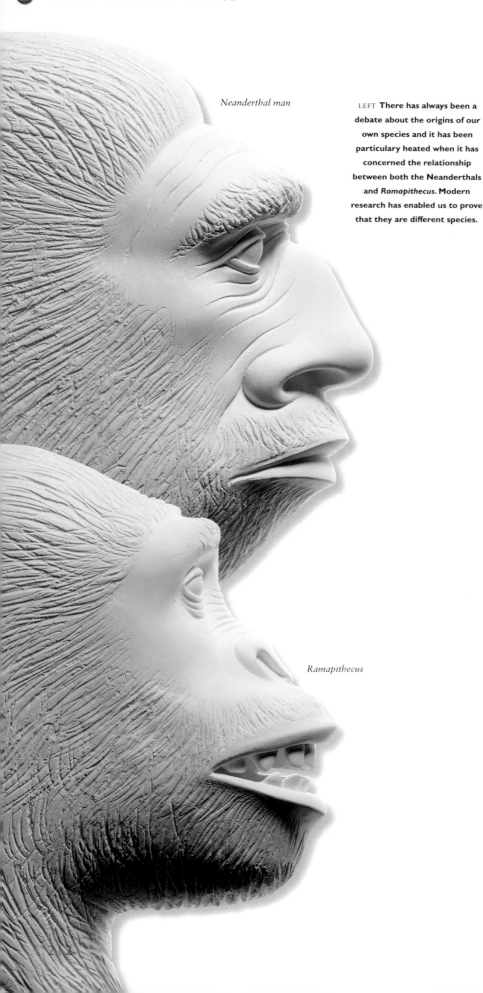

Neanderthal man

LEFT **There has always been a debate about the origins of our own species and it has been particulary heated when it has concerned the relationship between both the Neanderthals and *Ramapithecus*. Modern research has enabled us to prove that they are different species.**

Ramapithecus

WHO WERE THE NEANDERTHALS?

The Neanderthals were a distinct human population that lived in Europe and Asia – but not Africa – from about 100,000 to 30,000 years ago. They had large brains, but their skulls were flatter and more elongated than those of modern humans, and they had larger brow-ridges. Their short stocky bodies were very muscular and well-adapted to the conditions of the European ice age. They made stone tools, took care of the sick and disabled, and buried their dead.

A much-mooted issue during the past century has been the relationship of the Neanderthals to our species, *Homo sapiens*. Some anthropologists believe they were merely an anatomically distinct subspecies, *Homo sapiens neanderthalensis*, but others regard them as an entirely separate species, *Homo neanderthalensis*. Each interpretation has relied on the same fossils, and until recently there seemed to be no definitive way of settling the issue, one way or the other.

In 1997 Svante Pääbo and associates applied the techniques of molecular palaeontology to the question of Neanderthal relationships. They extracted mitochondrial DNA from the bones of the original Neanderthal specimens discovered in the Neander Valley of Germany in 1857, and compared these mtDNA sequences with those of living humans and apes. The difference between Neanderthal mtDNA and modern human DNA was much greater than the differences between current human geographical variants, but the Neanderthal's was much closer to human than chimpanzee mtDNA.

From these data it was calculated that the Neanderthal lineage split from the lineage leading to modern humans about 500,000 years ago. Since all modern humans have an ancestor that lived about 140,000 years ago, the Neanderthals – based on this one mtDNA sequence – clearly were a different species. So analysis of fossil molecules has helped to resolve another long-standing dispute about how an extinct species is related to a species that is still living – in this case our own, *Homo sapiens*, 'wise man'.

AN AFRICAN EVE?

When and where did our own species, *Homo sapiens*, arise? Molecular studies, and in particular the study of mitochondrial DNA, have played a key role in resolving the debate between competing hypotheses. Because mtDNA evolves much more rapidly than nuclear DNA it serves to measure the variation between populations of living humans, and also provides a way of dating their separation.

The debate over the origins of modern humans had thrown up two conflicting theories. One, the 'multiregional hypothesis', saw continuity between the founding populations of *Homo erectus* and *Homo sapiens* in each geographical region. This model depends upon a process of parallel evolution occurring on three continents over a period of at least half a million years. Its proponents find support for their thesis in supposed anatomical resemblances between living populations and their presumed fossil antecedents. In short, they believe that African *Homo erectus* evolved into modern Africans, Asian *Homo erectus* into modern Asians and European *Homo erectus* into modern Europeans.

The 'recent replacement theory' was a minority opinion, advocated particularly by British palaeoanthropologist Christopher Stringer. It hypothesizes the evolution of *Homo sapiens* in a single location, most likely sub-Saharan Africa, on the basis of hominid remains that closely resemble the skulls of modern *Homo sapiens*. This population was thought to have moved rapidly out of Africa throughout the Old World and caused the extinction of the resident populations of *Homo erectus*.

In 1987 the recent replacement model was strongly supported by comparing the mitochondrial DNA of native Africans, Asians, Caucasians, Australians and New Guineans. The findings were striking in three respects. First, the variability observed was greatest by far in Africans, implying that the African population is the oldest and is ancestral to the Asians and Caucasians. Second, the human mtDNAs showed

BELOW **Our own species, *Homo sapiens*, is the only survivor of several *Homo* species that have evolved over time. Molecular evidence implies that we originated in Africa about 140,000 years ago, spread out across the world and somehow eliminated competitors such as *Homo erectus* and Neanderthal man.**

Human

The African Eve was one of a flourishing population, but her genetic legacy is the only one that has survived

The heritage of every human alive on the planet lies on the savanna grasslands of tropical Africa

Mitochondrial DNA is passed on through the female line, since it is not carried by sperm cells

relatively little variability compared to ape mtDNA, with only about one-tenth as much as that of chimpanzees and orang-utans. This implied that our species has a recent origin.

The time factor was estimated by comparing modern human mtDNA with chimpanzee mtDNA and applying a molecular clock. The variability in African mtDNAs was one twenty-fifth as much as the average difference between human and chimpanzee mtDNA. One twenty-fifth of the five-million-year human-chimpanzee divergence is 200,000 years. Therefore the mitochondrial mother of us all, from whom all modern mtDNAs have descended with slight modifications, must have lived in Africa about 200,000 years ago. Inevitably this unknown female forebear has been dubbed the Mitochondrial Eve.

The fact that all modern mtDNAs originated from one woman does not mean that only one woman was alive at the time of origin. It is likely that 'Eve' was one of several thousand early members of our species, but her mtDNA is the only one that has survived to the present day. MtDNA lineages are lost when women have no offspring, or when they have only male offspring

ABOVE **The fast, yet steady mutation rate of mitochondrial DNA has enabled molecular biologists to trace our ancestry back through the female line to an 'African Eve'. She may have lived between 140,000 and 200,000 years ago – a mere instant in geological time compared to the four-billion-year history of life on Earth.**

who cannot pass their mtDNA on to the next generation. These losses are repeated generation after generation until only one of the original lineages remains.

Like mtDNA in women, a portion of the male Y-chromosome – the male sex chromosome – does not recombine but is passed unaltered through the male lineage. Almost no variation is observed in this chromosome segment in males alive today, which implies that the Y-chromosome lineage originated relatively recently in one individual. Naturally this hypothetical founder male has been called Adam. Y-chromosome analysis places Adam in Africa 100,000-200,000 years ago, and the latest estimate for the origin of *Homo sapiens*, based on sequences from the entire mitochondrial genome, is more precise at about 140,000 years.

More than a dozen independent studies of human DNA, proteins, and gene frequencies confirm the original mtDNA evidence that our species arose in Africa very recently in geological time. All of which goes to show that in the final analysis – although nothing is final in the study of evolution – Darwin was right.

AT A GLANCE

1856

Gregor Mendel discovers the basic statistical laws of heredity through his work on the edible pea.

1858

Charles Darwin and Alfred Wallace independently propose the theory of the evolution of species by natural selection.

ALFRED WALLACE

1909

Thomas Hunt Morgan shows that inheritance of white eyes in fruit flies is sex-linked.

INDIVIDUALITY

1954

George Gamow suggests that the nucleotide bases on DNA might form a genetic code.

1956

George Palade discovers the ribosome, where RNA is used to synthesize protein.

1960

Marshall Nirenberg and J. Heinrich Matthaei unlock the genetic code of bases.

1961

François Jacob and Jacques Monod, with André Lwoff, discover regulator genes.

1963

C. R. Woese discovers that the rRNA of the bacteria-like archaea is very different from that of normal bacteria.

WHALES AND HIPPOPOTAMI

1964

Robert Holley analyzes mRNA for the amino acid alanine, the first naturally occurring nucleic acid to be completely analyzed.

1965

Emile Zuckerkandl and Linus Pauling notice that the number of amino acids present in haemoglobin sequences is proportional to the time each species has been diverging from a common ancestor – so providing a molecular clock.

THE GREAT DEBATE

1967

Nirenberg and Har Khorana complete the genetic code dictionary.

1972

Charles Sibley and Jon Ahlquist find that many similar-looking groups of birds have very different DNA, showing that they are similar because they have evolved similar adaptations, and not because they have a common ancestor.

EVOLUTION

1971

Daniel Nathans and Hamilton Smith use restriction enzymes to snip DNA into separate strands.

1972

Paul Berg splices together DNA strands snipped with restriction enzymes. This is called recombinant DNA, and is vital for genetic engineering.

THE ORIGIN OF THE SPECIES

1982

Vincent Sarich and Allan Wilson conclude from studies of their albumins that humans, chimpanzees and gorillas had a common ancestor five million years ago.

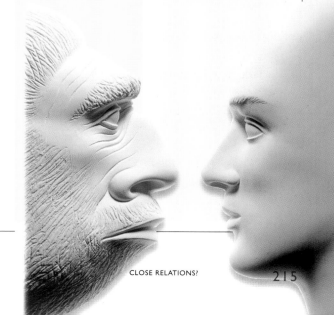

CLOSE RELATIONS?

DNA MAN

1952

Alfred Hershey and Martha Chase prove that DNA carries genetic information.

1953

Francis Crick and James Watson show that DNA has a double-helix structure.

Glossary

Absolute zero – Zero kelvin (-273°C) The lowest temperature that any substance can approach, when its constituent atoms are virtually stationary.

Accretion tectonics – The science of continental assembly from fragments of continental crust.

Acid – A substance that dissolves in water and yields hydrogen ions.

Alchemy – A primitive form of chemistry, based on magic rather than science.

Algebra – Branch of mathematics in which symbols are used to represent unknown numbers.

Alkali – A solution of a base in water, containing hydroxide ions that neutralize the hydrogen ions of an acid.

Alkali metal – Highly reactive metals in Group I of the periodic table.

Amino acid – A compound containing an amino group of two hydrogen atoms linked to a nitrogen atom; 20 naturally-occurring amino acids are the basis of all proteins.

Antibody – A blood protein that neutralizes foreign bodies such as pathogens.

Archaea – Primitive single-celled prokaryotic organisms resembling bacteria.

Atom – The smallest particle of an element that is characteristic of that element. Each element has its own distinctive atomic structure.

Atomic mass – The total number of protons and neutrons in the atomic nucleus.

Atomic nucleus – The point where all the mass of the atom is concentrated, within a cloud of electrons.

Atomic number – The number of protons in the atomic nucleus.

Australopithecine – An 'ape-man' of the genus Australopithecus, which preceded humans of the genus *Homo*.

Bacteria – Single-celled prokaryotic organisms.

Base – A substance that dissolves in water to form an alkali.

Big Bang – The event in which space, time and matter are thought to have been created, some 16 billion years ago.

Big Crunch – A possible future reversal of the Big Bang.

Black dwarf – A dead star that has exhausted all its energy.

Black hole – The collapsed remains of a giant star, so dense that its gravity prevents the escape of light.

Calculus – Branch of mathematics that calculates immediate rates of change.

Carbohydrate – A sugar or sugar-based polymer such as starch or cellulose.

Catalyst – A substance that speeds up a chemical reaction without being consumed itself.

Catenation – The linking together of carbon atoms to form chains, as in polymers.

Cells – The basic units from which living things are built.

Chaos theory – Branch of mathematics that deals with chaotic or unpredictable systems.

Chemical bond – An electrostatic bond between atoms.

Chemical compound – A substance consisting of more than one element, with the different atoms bound together in a fixed ratio.

Chirality – The ability of a molecule to exist in two mirror-image forms, with different properties.

Chloroplasts – The structures in plant cells that carry out photosynthesis.

Chromosome – Structures within the cell that are made partly of DNA, and carry hereditary information.

Codon – A sequence of three bases in RNA that instructs a ribosome to build a particular amino acid into a protein.

Colour charge – The force that holds quarks together.

Complex numbers – A number written in the form $x + iy$, where x and y are real numbers and i is the square root of -1.

Continental drift – The theory that the continents move across the surface of the Earth.

Covalent compound – A compound formed of discrete molecules.

Crystal – A natural geometric solid, with a shape that reflects its atomic structure.

Cytoplasm – The cell contents surrounding the nucleus.

DNA – Deoxyribonucleic acid, a natural polymer that carries the genetic information necessary to build an organism.

Dominant character – An inherited trait that is always expressed.

Doppler shift – The 'police siren' effect, when waves propagated by a moving body have their wavelength stretched as the body moves away.

Electromagnetic field – The force field around an electromagnet.

Electromagnetic radiation – Waveforms in the electromagnetic field that, depending on their wavelength, take the form of radio waves, microwaves, infra-red, visible light, ultraviolet, X-rays and gamma rays.

Electron – One of the cloud of negatively-charged particles that surrounds the atomic nucleus.

Electron shell – A group of electrons with the same energy level, occupying a particular 'orbit' around the nucleus.

Element – A substance consisting of identical atoms (or, strictly, atoms with the same atomic number).

Entropy – The degree of disorder into which all systems decline, inevitably.

Enzyme – A protein that acts as a catalyst for a biochemical reaction.

Equation – An expression that represents the equality of two expressions involving some unknown quantity x.

Eukaryote – An organism whose cells contain nuclei.

Evolution – A change in the heritable characteristics of an organism over the generations.

Fault – A fracture in the Earth's crust that has allowed movement.

Fibonacci numbers – An infinite sequence of numbers in which each number is the sum of the previous two numbers: 1, 1, 2, 3, 5, 8, etc.

Fossil – The remains of an ancient animal or plant, preserved in the rock.

Fossil record – The record of past life preserved in fossil form.

Galaxy – A vast gravity-bound system of stars, gas and dust, often with a spiral form.

Gamete – A reproductive cell, either a sperm (male) or ovum (female).

Gene – A unit of inheritance, formed from a length of DNA.

Genetic code – The sequence of bases in RNA (and DNA) that acts as a code for the construction of proteins.

Genome – The complete collection of genes in a cell.

Genus – An artificial grouping of apparently allied species.

Geometry – Branch of mathematics concerned with the properties of shapes.

Gluons – The quanta of the colour charge that binds quarks into nucleons.

Gravity – The attraction between objects that have mass.

Homeotic genes (hox genes) – The genes that determine development and body plan.

Hominid – Any member of the primate family Hominidae, including humans and australopithecine 'ape-men'.

Hydrocarbon – A compound such as methane that contains only hydrogen and carbon atoms.

Igneous rock – A rock formed by the cooling and crystallization of molten magma.

Inertia – The natural tendency of any object with mass to continue moving at the same velocity unless subjected to a force.

Ion – An atom or molecule that has lost or gained one or more electrons, so it has a positive or negative charge.

Ionic compound – A substance whose component atoms are held together in an extended lattice by electrostatic forces.

Isotope – Different isotopes of an element have different numbers of neutrons in their atomic nuclei.

Kinetic – To do with movement.

Lava – Molten rock erupted at the surface of the Earth.

Logarithm – The exponent of a number indicating the power to which a fixed number, the base (usually 10), must be raised to obtain a given number.

Magma – Molten rock beneath the surface of the Earth.

Magnetosphere – The envelope around the Earth formed by its magnetic field.

Mantle – The 2,900-km-deep layer of the Earth's interior, extending from the crust to the core.

Glossary

Mass – The property of an object that determines its weight under the influence of gravity. In reduced gravity it weighs less, but its mass is unaffected.

Meiosis – The type of cell division that creates reproductive cells, in which genetic material is shuffled by recombination.

Metal – A substance in which positive ions are bonded together by a pool of negative electrons.

Metamorphic rock – Rock that has been altered by pressure and/or heat.

Metazoa – Multi-celled animals.

Mineral – A naturally-occurring element or compound with an ordered internal structure.

Mitochondria – The structure within cells that turn sugars into energy.

Mitochondrial DNA (mtDNA) – DNA held within the mitochondria, which evolves at a rapid but steady rate over time.

Mitosis – The type of cell division that creates two identical cells.

Möbius band – A one-sided continuous surface, formed by giving a half twist (through 180°) and joining the ends together.

Molecular clock – The steady accumulation of inherited mutations in DNA that can be used to put dates on evolutionary events.

Molecule – A group of atoms bound together into a discrete, well-defined unit.

Monomer – A compound whose molecules can be chained together into a polymer.

Morphology – The structure and appearance of an organism.

Mutation – A change in an inherited DNA sequence caused by a 'mistake' in code transcription.

Nanotechnology – Technology on the molecular scale.

Natural selection – The theory that species evolve by 'survival of the fittest'.

Nebula – A diffuse interstellar cloud of gases and dust.

Neutron – A particle in the atomic nucleus that has no charge.

Noble gases – The group of very unreactive gases in group 0 of the periodic table.

Nuclear fission – The splitting of the atomic nucleus to form two smaller nuclei, releasing energy.

Nuclear fusion – The fusion of two atomic nuclei into one larger nucleus, releasing energy.

Nucleic acid – Either DNA or RNA.

Nucleon – A proton or neutron.

Nucleotide – One of the components of a nucleic acid.

Number theory – The study of whole numbers (integers).

Organic chemistry – The chemistry of carbon compounds.

Organism – Any living thing.

Paleomagnetism – A record of the Earth's magnetic field preserved in ancient rocks at the time of their formation.

Photon – A particle or quantum of electromagnetic radiation.

Plasma – A gas-like substance in which the atoms are dissociated into electrons and ions.

Plate tectonics – The study of large-scale movements of the Earth's crust.

Polarized light – Light that vibrates in only one plane.

Polymer – A long-chain molecule formed by linking many smaller molecules.

Prime number – An integer that only be divided by itself or 1, such as 2, 3, 5, 7 and 11.

Prokaryote – A single-celled organism with no distinct nucleus.

Protein – A natural polymer of amino acids that coils into a distinctive structure.

Protein immunology – The use of antibodies to determine the relationships of different species. Antibodies specific to one species have variable reactions with others depending on their genetic similarity.

Proton – A particle in the atomic nucleus that has positive charge.

Pythagoras' theorem – A theorem in geometry stating in a right-angled triangle the sum of the length of the hypotenuse (the side that is opposite the right angle) equals the sum of the squares of the other two sides.

Quantum – A unit of matter or electromagnetic radiation, usually visualized as a particle.

Quantum electrodynamics – A theory that describes both the wave and particle behaviour of electromagnetic radiation.

Quantum mechanics – A theory describing the behaviour of matter, usually on the atomic scale.

Quarks – The particles that, in groups of three, form protons and neutrons.

Radioactivity – The emission of particles from atomic nuclei in the process of radioactive decay.

Radiometric dating – Finding a rock's age using the decay rate of radioactive elements in the sample.

Recessive character – An inherited trait that is only expressed if it is not suppressed by a dominant character.

Recombination – The shuffling of genetic information during the creation of reproductive cells that makes offspring different from their parents.

Red giant – An expanded, giant star that appears cool and red.

Redshift – A Doppler shift towards red in the light from a distant star, caused by its motion away from Earth.

Reflection – The bouncing of a wave off a surface or boundary.

Refraction – The bending of a wave as it passes across a boundary, owing to the difference in wave speed in the two materials.

Ribosomes – The sites of protein synthesis in the cell.

RNA – Ribonucleic acid, the natural polymer that transcribes genetic information from DNA and delivers it to the sites of protein synthesis.

Sedimentary rock – Rock formed from compacted, cemented sediments.

Solar wind – A plasma of charged particles emitted by the Sun, which shapes the Earth's magnetosphere.

Spacetime – The four-dimensional continuum of time and space, which is warped by gravity.

Species – Distinct individual type of organism. Organisms of the same species naturally interbreed.

Spectrum – Sequence of waves of progressively changing wavelength. A rainbow is a spectrum of visible light.

Spreading centre – Site where oceanic crust is being formed.

Strata – A layer of sedimentary rock.

Stratigraphic record – Record of geological events preserved in the rock strata.

String theory – A concept in which all the phenomena of quantum physics can be explained in terms of vibrating loops of 'string'.

Subduction zone – Inclined zone where one plate of the Earth's crust is plunging beneath another.

Supernova – An exploding star.

Taxonomy – The science of classifying living and extinct organisms.

Terrane – A fragment of continental crust that has become attached to other fragments or continents with different origins.

Thermodynamics – The study of heat flow.

Topology – Branch of mathematics that is concerned with surface properties that do not change under distortion.

Trigonometry – Branch of mathematics that deals with the sides and angles of triangles.

Valency – A figure that describes the number of hydrogen atoms that an atom of any element may combine with.

Vertebrate – An animal with a backbone and internal skeleton.

Velocity – The rate of change of position with time, both in terms of speed and direction.

Wavelength – The distance between two identical points in a wave pattern, such as two wave peaks.

Wave-particle duality – Paradoxical behaviour of light as both a wave and a particle, which applies to all forms of electromagnetic radiation as well as matter.

Weak nuclear force – The force responsible for radioactive decay.

White dwarf – Small, hot star formed from a collapsed red giant.

Whole number – A number that can be expressed as the sum or difference of a finite number of units, as 1, 2, 3, etc., also known as an integer.

Zygote – Fertilized reproductive cell, formed by the union of sperm and ovum.

Notes on Contributors

DAVID BAILEY is a freelance science writer based in Cambridge. Specializing in chemical sciences, he has written on these subjects for the British popular media, *New Scientist*, *Science*, *Encarta* CD-ROM and countless specialist magazines. He is Contributing Editor for the journal *Analytical Chemistry* and a weekly columnist for the *Alchemist* magazine. He also publishes a webzine – *Elemental Discoveries*. He was the *Daily Telegraph* Young Science Writer of the Year in 1992, and 1995 runner-up in the Chemical Industries Association Awards.

RICHARD DAWKINS holds Oxford's newly endowed Charles Simonyi Chair of Public Understanding of Science and is also a Professorial Fellow of New College, Oxford. He has written many highly successful books, including: *The Selfish Gene*; *The Extended Phenotype*; *The Blind Watchmaker*; *River Out of Eden*; and *Climbing Mount Improbable*. A frequent guest on British television and radio, he gave the Royal Institution Christmas Lectures in 1991 and the Dimbleby Lecture in 1996. His prizes and awards include the Royal Society of Literature Prize, the *Los Angeles Times* Literary Prize, the Silver Medal of the Zoological Society of London, the Michael Faraday Award of the Royal Society of London, Nakayama Prize for Human Science and the International Cosmos Prize, Osaka, Japan.

NICK FLOWERS is a graduate of the International Space University and works at the Mullard Space Science Laboratory, University College, London, where he has studied for his PhD in space physics. He won the *Daily Telegraph* Young Science Writer of the Year Award in 1995 and has written space science features for newspapers and science magazines ever since.

JOHN GRIBBIN is a Visiting Fellow in Astronomy at the University of Sussex. He is the author of more than eighty non-fiction books and several science fiction novels, and is best known for his writings on quantum physics, *In Search of Schrödinger's Cat*, *Schrödinger's Kittens*, and the biography of Richard Feynman that he wrote with Mary Gribbin.

JEROLD M. LOWENSTEIN is Clinical Professor of Medicine at the University of California, San Francisco, and Chairman of the Department of Nuclear Medicine at California Pacific Medical Center. He does research in molecular evolution and was a pioneer in the detection and analysis of fossil molecules, work which provided background for the novel and movie *Jurassic Park*. He has written numerous articles on biology, medicine, and evolution for scientific and popular publications.

IAN STEWART is Professor of Mathematics at the University of Warwick. He has written over sixty books including: *The Magical Maze*; *Figments of Reality*; *Nature's Numbers*; *The Collapse of Chaos*, *Does God Play Dice?*; *From Here to Infinity*; and *Fearful Symmetry*. In 1995 he was awarded the Royal Society's Faraday Medal for furthering the public understanding of science. He delivered the 1997 Royal Institution Christmas Lectures on BBC television.

PETER WARD is Professor of Geological Sciences, Professor of Zoology, and Curator of Palaeontology at the University of Washington, Seattle. He is the author of several books, including: *On Methuselah's Trail*; *The End of Evolution*; *The Call of Distant Mammoths – What Killed the Ice Age Mammals*; and *Time Machines*. His work has been profiled in *National Geographic*, the *New York Times*, *Science* and the *New Scientist*, among others, and he frequently appears in the media.

ADRIENNE ZIHLMAN is Professor of Anthropology at the University of California, Santa Cruz. Her research on human evolution has involved many field trips to Africa, where she has studied human fossils and observed chimpanzees, gorillas, and monkeys. Her controversial proposal that the little-known pygmy chimpanzee (*Pan paniscus*) is the living species most like the common ancestor of humans and chimpanzees has been supported by the most recently discovered human fossils, which are more than four million years old. She has also written extensively on the role of women in evolution and is author of the widely-used textbook *The Human Evolution Colouring Book*.

Acknowledgements

Archiv für Kunst und Geschichte, London: pp.8/90, 12ML, 12B (Swiss History of Pharmacy, Basel), 13BR, 14T, 51BR, 56TR, 56B, 70T, 79, 80, 92B, 169, 192, 193, 215 (Down House, Kent).

Orsi Battaglia/AKG, London: pp.26T, 49BL (Museo dell Opera del Duomo, Florence). Eric Lessing/AKG, London: 12MR (Church of St John, Torun), 51TR (Amlas Castles Coll., Innsbruck).

Bridgeman Art Library: pp.12BL, 14ML, 54BR, 152.

e.t.archive: pp.65TR, 87TR.

Natural History Museum, London: pp.206/7.

NASA: pp.89T, 90TR, 91, 104T, 105T, 108, 110BL, 110TR, 113, 114, 115.

Science Photo Library: cover pictures (all), pp.10, 14T, 15T, 15M, 15B, 17B, 30BL, 34, 40T, 41B, 45T, 46/7, 47, 48T, 48B, 52/3, 58TL, 58TR, 60, 62, 72T, 73TL, 73BL, 73R, 74T, 76, 81T, 87B, 88TR, 92TL, 93TR, 95T, 96, 99, 102/3, 107T, 107BR, 112, 115, 117B, 121MR, 122TL, 129T, 129R, 130, 131, 134L, 134TR, 135, 139, 140, 141, 143, 144T, 147R, 149TR, 151, 156, 159BR, 166/7, 177, 179TR, 181B, 185BR, 191TR, 197TR, 205T, 205B, 210TR, 211L.

Stock Market, London: pp.18/19, 20, 35TR, 44B, 59BL, 64/5, 74B, 84B, 108R, 149TR, 150/1, 155, 158TL, 175, 199, 204.

Tony Stone: p.195.

Index

Index